Ground-Based Microwave Radiometry and Remote Sensing

Methods and Applications

Ground-Based Microwave Radiometry and Remote Sensing

Methods and Applications

Pranab Kumar Karmakar

CRC Press
Taylor & Francis Group
Boca Raton London New York

CRC Press is an imprint of the
Taylor & Francis Group, an **informa** business

CRC Press
Taylor & Francis Group
6000 Broken Sound Parkway NW, Suite 300
Boca Raton, FL 33487-2742

First issued in paperback 2017

© 2014 by Taylor & Francis Group, LLC
CRC Press is an imprint of Taylor & Francis Group, an Informa business

No claim to original U.S. Government works

Version Date: 20130916

ISBN 13: 978-1-4665-1631-1 (hbk)
ISBN 13: 978-1-138-07452-1 (pbk)

Library of Congress Cataloging-in-Publication Data

Karmakar, Pranab Kumar.
  Ground-based microwave radiometry and remote sensing : methods and applications / Pranab Kumar Karmakar.
      pages cm
    ISBN 978-1-4665-1631-1 (hardback)
      1. Atmospherics. 2. Atmosphere--Research. 3. Radio meteorology. 4. Radiation--Measurement. 5. Microwave measurements. I. Title.

  QC973.4.A85K37 2013
  551.5028'7--dc23   2013026465

Visit the Taylor & Francis Web site at
http://www.taylorandfrancis.com

and the CRC Press Web site at
http://www.crcpress.com

Dedicated to the Lotus Feet of Lord Jagannath

Contents

Preface

Remote sensing by using microwaves has become an important diagnostic tool for probing the atmosphere and surface of planetary objects. The term *microwave remote sensing* encompasses the physics of microwave propagation and its interaction with atmospheric ambient particles.

The basic components of microwave remote sensing are the sensor–scene interaction, sensor design, and its application in geosciences. This book is mainly for physicists and engineers working in the area of microwave sensing of the atmosphere; it is not for ultimate users like geologists and hydrologists. An attempt has been made to establish a link between the microwave-sensor response and ambient atmospheric thermodynamic parameters, like water vapor content, temperature, nonprecipitable cloud-liquid-water content, and rain in the tropical, temperate, and polar regions. It should be mentioned here that of several types of sensors, such as radar, radiometer, LIDAR, etc., we have described the ground-based radiometric application in remote sensing of the atmosphere, which in a sense may be called microwave radiometry.

Radiosonde observations (RAOBs) are considered to be the most fundamental and acceptable method for atmospheric temperature and water vapor measurements and profiling, despite their inaccuracies, cost, sparse temporal sampling, and logistical difficulties. A better technology has been sought for the past few decades, but until now, no accurate continuous all-weather technology for probing the atmosphere has been demonstrated. Laser radars (LIDARs) and Fourier transform infrared spectrometers can profile temperature and water vapor, but not in the presence of clouds. The only reasonable and acceptable solution is the highly stable frequency-agile radiometric temperature and water vapor measurements. This radiometric method gives us continuous unattended measurements. It also has the capability to profile cloud liquid water, a capability absent in RAOBs and all other systems except for in situ aircraft devices.

The term *microwave* does not have a very precise definition, but the accepted division of the electromagnetic spectrum is 3–30 GHz. The technology that is being used now offers large application potential in this band. The measurement techniques divide according to the wavelength sensitivity of the detecting apparatus. For example, one of the principal advantages of using microwave systems is their all-weather capability. The principal disadvantage of microwave systems is their sensitivity to propagation path effects that are induced by water vapor. But on the other hand, this becomes a boon to remote-sensing scientists as far as measurement of water vapor content or distribution is concerned. It is impossible to model water vapor with a high degree of accuracy. But still, the most promising technique available for making vapor measurements is the use of passive microwave radiometry. This is the reason for this book based on microwave radiometric measurements and their applications.

This book consists of eight chapters. The first chapter deals with the fundamentals of microwave remote sensing. In it we discuss the atmospheric influences on the electromagnetic spectrum. At a given frequency, the atmospheric absorption coefficient is a function of three basic parameters: temperature, pressure, and water vapor density. Water vapor absorption essentially depends on water vapor density, and likewise, oxygen absorption is dependent on pressure and temperature. Since water vapor density and pressure decrease exponentially with increasing altitude, the major contribution in finding the zenith opacity (dB) is provided by the layers closest to the surface. An attempt has been made to find out the window frequency lying between the two maxima occurring at 60 and 120 GHz. These two maxima are due to the atmospheric oxygen and are called the "oxygen maxima" in the microwave spectrum. For this purpose, eight places have been chosen, four from the northern latitude and four from the southern latitude. The window frequencies obtained over the northern and southern latitudes are described in detail.

The second chapter deals with the basic laws of radiation and its fundamentals. Included in this chapter are the basic principles and terminology used in microwave radiometry. It deals with the measurement of incoherent electromagnetic radiation from an object obeying the laws of radiation fundamentals. A microwave radiometer is capable of measuring very low-level microwave radiation from an object. Sometimes it is possible to establish a useful relationship between the magnitude of the radiation intensity and a specific terrestrial or atmospheric parameter of interest. Once the relationship is established, the desired parameter can thus be obtained from microwave radiometer measurements. The potential use of a ground-based microwave radiometer may be considered for atmospheric modeling purposes. But from the modeling point of view, the approach to ambient water vapor, temperature, and rainfall signature requires thorough insight into the electromagnetic interaction between microwave radiation and the medium concerned, since the radiometric response depends on various radiative sources while radio waves propagate through the atmosphere. All these are described in detail, along with different gaseous and cloud models that are in use worldwide.

Water vapor content produces a sizable attenuation in the microwave water vapor band, i.e., 20–30 GHz. At the same, it is worthwhile to mention that its spatial distribution is not uniform, and hence it needs special attention for local measurement. Microwave radiometry is the only solution for its continuous and unattended monitoring. Initially it was assumed that the radiometer possesses radio visibility extended to infinity, but this may not be the case in a place where water vapor seems to be a key factor in contributing toward absorption in the microwave band. But, on the other hand, the oxygen spectra ranging from 50–70 GHz seem to be more complex and attenuation is sizable. Hence, special attention should be given to this band, which is generally used as a temperature profiler. The zenith-looking radio visibility, i.e., the height limit up to which there is no substantial increase in attenuation in the zenith direction, also needs to be explored. The radio visibility is defined here as the height up to which the variation of total attenuation is less than or equal to 1% of that of the immediate preceding height. Conversely, we can say that 99% of a radio signal can reach the height where the attenuation is equal to, or less than 1%. For the sake of calculation, one should consider the thickness of each slab equal to 10 m or less. But on the other hand, while defining this radio visibility or height limit in the oxygen band, the slab thickness is considered equal to 10 m.

The percentage change in attenuation is considered to be 0.1%, despite the fact that the attenuation is high in the oxygen band compared to that in the water vapor band. These height limits are well studied in both the water vapor band and the oxygen band for a zenith-looking radiometer in Chapter 3.

Chapter 4 encompasses radiometric sensing of water vapor, temperature, and nonprecipitable cloud liquid water. Water vapor is perhaps the most important minor constituent that can affect the thermodynamic balance, photochemistry of the atmosphere, sun–weather relationship, and biosphere. The vertical and horizontal distribution of water vapor, as well as its temporal variation, is essential for probing the mysteries of several effects. In fact, this triggered the need for the measurement of ambient water vapor for numerical weather predictions, short-term as well as severe storm forecasting. On the other hand, liquid water content measurement of clouds also provides very important input to the global circulation model (GCM). In fact, the electromagnetic radiation at microwave frequencies interacts with the suspended atmospheric molecules, in particular with oxygen and water vapor. This interaction may be manifested in two ways in terms of complex refractivity. The imaginary part is generally expressed as attenuation, to be discussed in Chapter 4, and the real part deals with propagation delay, which will be discussed later in Chapter 7. For this purpose, the radiative transfer equation is dealt with in terms of various models, like the forward model and inverse model. This inverse model has been categorized into several types. Accordingly, its advantages and disadvantages are elaborated in Chapter 4 from an application point of view. It is well accepted that multifrequency radiometric sensing is essential to get an acceptable value of the thermodynamic variables. But the higher the frequency, the less the possibility of getting plentiful data will be. For example, an extensive amount of data in the water vapor band may be available all over the world. But, installation of another window frequency in the higher range, for example, 90 GHz, is costly, and simultaneity may not be possible to maintain. In that case, it is of interest to examine between-channel predictability. Such considerations are important when attenuation is being estimated at various locations. For this purpose, a case study in Denver, Colorado, and San Nicolas Island, California, on channel predictability at 20, 34, and 90 GHz has been discussed. Another important case study for thermodynamic profiling has been incorporated at Boulder, Colorado, and Huntsville, Alabama, in great detail. Some examples of radiometric retrievals from a variety of dynamic weather phenomena, including upslope super-cooled fog, snowfall, a complex cold front, a nocturnal bore, and a squall line accompanied by rapid variations in low-level water vapor and temperature, are presented. Included also are five different constrained examples derived from the radiometric measurements.

Chapter 5 describes the measurement technique of water vapor in the polar region. Several campaigns have been organized by the European community in the northern hemisphere to understand the ozone characteristics within polar vortices.

The Antarctica Microwave Stratospheric and Tropospheric Radiometers (HAMSTRAD) program aims to develop two ground-based microwave radiometers to sound tropospheric and stratospheric H_2O above Dome C (Concordia Station), Antarctica (75°S, 123°E) over a long time period. HAMSTRAD-Tropo is a 183 GHz radiometer for measuring tropospheric H_2O. The Arctic Winter Radiometric

Experiment was held at the Atmospheric Radiation Measurement (ARM) Program's North Slope of Alaska (NSA) site near Barrow, Alaska. During this experiment, 25 channels were deployed. Among these, 12 channels were selected at the lower wing of the oxygen line and 2 channels at 89 GHz (one vertical and the other horizontal). Seven channels were distributed around 183 GHz, two polarized channels at 340 GHz (H and V), and three channels around the 380.2 GHz water vapor line. These channels were selected to provide simultaneous retrievals of precipitable water vapor (PWV) and liquid water path (LWP) and low-resolution temperature and humidity profiles.

All of these measurement techniques, such as the one-dimensional variation technique and its advantages and disadvantages in both the Arctic and Antarctic polar regions, are discussed in detail in Chapter 5.

Chapter 6 includes detailed studies of the measurement of integrated water vapor content by deploying a microwave radiometer. For this purpose, initially a single channel at 22.234 GHz was used with some simplifying assumptions while developing the algorithm for the measurement. There was some significant error found in comparison to the RAOB's result. But when radiometric outputs were used in a cloudless condition, the error was minimized. So, this tells us the need for separating the cloud liquid from water vapor by deploying dual-frequency channels and developing the proper algorithm. In this context, the frequency selection criteria are very crucial. Hence, in this book, different frequency pairs have been chosen for this purpose, and the improvement of the results are described sequentially. Finally, a particular pair is prescribed for the measurement, along with the model for integrated water vapor content over a particular place of choice.

Chapter 7 deals with radiometric measurement of propagation path delay, which is of paramount interest as far as the geodetic application is concerned. Atmospheric water vapor is a limiting source of error in determining the baselines by the technique of very long–baseline interferometry (VLBI). The use of global positioning systems (GPS) offers even more precise geodetic measurements. On the other hand, it may be even more limited by the variable wet path delay component due to the very presence of water vapor in the atmosphere. In fact, the atmosphere plays a key role for operational geodesy. Basically, there are two different possible approaches: (1) the integral atmospheric model, where the atmospheric effects are incorporated in some form in the adjustment process, and (2) the peripheral atmospheric model, where the observations are corrected for the atmospheric effects before they are entered into the adjustment process. Both the advantages and the disadvantages are well illustrated from the modeling point of view in this chapter. The development of the integral model starts with the appropriate equations derivable from the geodetic measurements. Since the atmosphere affects the measurements, geodesists are able to estimate those properties to which their observations are sensitive. Among these, the vertically integrated delay of the radio signal due to water vapor in the atmosphere creates a lot of attraction. Because of the importance of water vapor to meteorology, the prospect of a new, relatively inexpensive instrument to determine its spatial and temporal distribution should be welcomed, and hence is the use of the multifrequency radiometer. Although it does not have vertical resolution like the radiosonde, it has the advantage of high temporal resolution and is able to detect cloud liquid water. It has also been used by many scientists to show the effect of liquid water in the millimeter-wave spectrum at a tropical location. However, the presence of liquid water does little to change path delay. But for the present purpose, we need to discriminate the effect of liquid, however small it is, as we will be handling the tropical climate.

The effect of delay of the microwave signal while propagating through the atmosphere is due to two parts: one is due to dry delay, and hence can be modeled easily with the use of surface parameters. The term *delay* refers to change in path length due to change in the refractive index during the propagation of radio signals through the atmosphere duly constituted by several gases. Their combined refractive index is slightly greater than unity and gives rise to a decrease in signal velocity. This eventually increases the time taken for the signal to reach the receiving antenna. Hence, the wet path delay depends on the precipitable water vapor in the column of air through which the signal propagates. In this chapter wet path delay in terms of daily, monthly, and seasonal variation over a tropical location is well described. But, since the variation of water vapor in a temperate climate is of little significance to delay, it is not discussed.

The last chapter is concerned with the influences of rain in the microwave signal. Effort has been made to illustrate the rain model using radiometric measurement at different locations over the globe, which include tropical and temperate climates. In this chapter, several real-time pictures of radiometric and disdrometer measurements have been presented for the sake of clarity.

I confess that I have freely consulted existing textbooks and many published research articles from different journals of repute during the preparation of the manuscript to raise the power and ability of the graduate and undergraduate students and professionals as well. I am indebted to those eminent authors from different countries all over the world and the publishers of the journals.

I specially convey my sincere gratitude to Dr. Randolph "Stick" Ware, chief scientist and founder, Radiometrics Corporation (United States), for extending valuable suggestions while preparing the manuscript.

The enthusiastic support I received from Ashley Gasque to streamline my thoughts is noteworthy for undertaking this challenging project. The support extended by different sections of the publishing house, CRC Press (United States), is also noted, for which I remain thankful to all of them.

Last but not the least, I must mention my wife, Anupama Karmakar, for her continuous support and encouragement to undertake the project, and also my son, Anirban Karmakar, for extending his untiring effort in drawing the figures cited in this book. Otherwise, it would not have been possible for me to complete this project.

Dr. Pranab Kumar Karmakar

Institute of Radiophysics and Electronics
University of Calcutta
Kolkata, India

About the Author

Pranab Kumar Karmakar is currently pursuing research work principally in the area of modeling of integrated water vapor and liquid water in the ambient atmosphere. Ground-based microwave radiometric remote sensing is his specialty. This includes vertical profiling of thermodynamic variables. He is currently involved in research and teaching at the post-graduate level at the Institute of Radiophysics and Electronics, University of Calcutta in India.

Since joining the University of Calcutta in 1988, Dr. Karmakar has published noteworthy outcomes of his research of tropical locations in different international and national journals of repute. All these culminated in a book entitled *Microwave Propagation and Remote Sensing: Atmospheric Influences with Models and Applications* published by CRC Press in 2012 as well as this book.

Dr. Karmakar was awarded the International Young Scientist award of the URSI in 1990. He has also received the South–South Fellowship of The Third World Academy of Science (TWAS) in 1997. He has acted as a visiting scientist at the Remote Sensing Laboratory, University of Kansas; Centre for Space Sciences, China; and National Institute for Space Sciences, Brazil.

1 Ground-Based Remote Sensing

1.1 INTRODUCTION: DEFINITION OF REMOTE SENSING

Remote sensing is the science or technology for acquiring information, like identifying, classifying, and determining objects of the earth's surface and environment from air or space by means of electromagnetic waves. The physical information may be obtained through the analyses of data from the atmospheric objects collected by using remotely located sensors that are practically not in physical contact with those objects. Precisely, remote sensing may be termed a tool for retrieving the physical properties of the objects of the atmosphere through the measurement of their electromagnetic emission or absorption or scattering characteristics.

Remote sensing techniques are categorized by the fact that they provide the qualitative and quantitative features of the earth and its atmosphere without any material contact. This technology is subdivided into two categories: active sensing and passive sensing.

Sensing technology, irrespective of active and passive mode, includes:

1. Sensor technology for obtaining information on atmospheric objects. It may be divided into three parts:
   a. Satellite based
   b. Aircraft based
   c. Ground based
2. Data acquisition and dissemination

1.2 MICROWAVE REMOTE SENSING AND ITS APPLICATION

Radiosonde observations (RAOBs) are considered to be the most fundamental and acceptable method for atmospheric temperature and water vapor measurements and profiling, in spite of their inaccuracies, cost, sparse temporal sampling, and logistical difficulties. A better technology has been sought for the past few decades, but until now, no accurate continuous all-weather technology for probing the atmosphere has been demonstrated. Laser radars (LIDARs) and Fourier transform infrared spectrometers can profile temperature and water vapor, but not in the presence of clouds. The only reasonable and acceptable solution is the highly stable frequency-agile radiometric temperature and water vapor measurements. This radiometric method gives us continuous unattended measurements. It also has the capability to profile cloud liquid water, a capability absent in RAOBs and all other systems except for in situ aircraft devices. Applications for this passive radiometric profiling include weather forecasting and now-casting, detection of aircraft icing and other aviation-related meteorological hazards, determination of density profiles for artillery trajectory and sound propagation, refractivity profiles for radio-ducting prediction, corrections to radio astronomy, satellite positioning and global positioning system (GPS) measurements, atmospheric radiation flux studies, estimation and prediction of telecommunication link degradation, and measurement of water vapor densities as they affect hygroscopic aerosols and smokes (Solheim et al., 1998). But as the situation demands, sometimes the meteorological observations provide valuable information.

Existing meteorological observational systems use radiosondes launched every 12 hours over the land in question. This system does not provide adequate information even in meteorology. Now-casting is such an example (Askne and Westwater, 1986), e.g., forecasts for 0–12 hours ahead. On that timescale we may have important short-lived phenomena of interest, including front movements, buildup of convective clouds, etc. In this regard, it may be mentioned that the well-known product-moment statistical formula was employed by Sen et al. (1989) in order to find out the correlation of the temporal variation of the water vapor density at different heights in the range 0–10 km, averaged over 6, 12, and 24 hours, to the variation of the integrated water vapor content. The results obtained are shown in Figure 1.1. It was found that the correlation coefficient between integrated vapor content and the surface water vapor density is 0.65, and it attains a broad maximum of 0.95 at the heights of 2–2.5 km and drops to as low as about 0.4 for a height of 10 km for the averaged times of 12 and 24 hours. For the 6-hour periods of averaging, however, there was no noticeable correlation. Thus, the heights over a range of 0.5 km centered around 2 km are found to be the most significant, representing the diurnal variation pattern of the integrated water vapor content. A comparative study of the diurnal pattern of the integrated water vapor content and

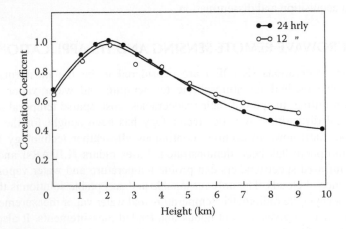

FIGURE 1.1 Correlation between the integrated water vapor content and vapor density at 10 km height; 240 cases have been considered. The correlation is a maximum at about 2 km height for 12- and 24-hourly averaging time.

water vapor density at 2 km height has been made between 24 hours and 48 hours, as shown in Figure 1.2. It is evident from this figure that a noticeable correlation exists in the case of 24-hourly averaging, but 48-hourly averaging of the same produces no appreciable correlation. The observed maximum correlation coefficient between the water vapor content at 2 km height and the integrated water vapor content suggests that the integrated water vapor content is not affected by the temporal variations of water vapor density at the surface or near 10 km height within the timescale of 12–24 hours. It is also noticed that for a time resolution between 12 and 24 hours the correlation is good, and therefore any transportation of vapor from the surface must have been a negligible effect within this timescale. Moreover, for a shorter timescale of 6 hours, again correlation is insignificant. This suggests the presence of finer and independent components of temporal variation at various heights. This situation can be well addressed by remote-sensing techniques using microwave radiometry.

Theoretical estimates of the microwave attenuation spectrum due to atmospheric gases indicate that under clear weather conditions, there exist minima around 30, 94, 140, and 220 GHz (Button and Wiltse, 1981). In between these windows there exist the maxima of attenuation due to water vapor and oxygen molecules (Rogers, 1953, 1954; Straiton and Tolbert, 1960; Van Vleck, 1947a, 1947b). The maxima due to

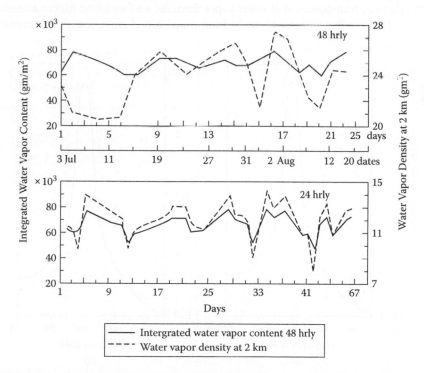

FIGURE 1.2 Temporal variation of the integrated water vapor content and vapor density at 2 km height. For 24-hourly averaging time, the correlation is good, but for 48-hourly averaging time, the correlation is not noticeable.

water vapor occur around 22.234 GHz and 183.311 and 325 GHz, while for oxygen the maxima occur around 60 and 118 GHz. The second-order water vapor cluster, i.e., dimer, has resonance lines at 212.1 and 636.6 GHz (Viktorova and Zhevakin, 1976). Zenith attenuation from 100 to 1000 GHz is discussed by Emery and Zavody (1979). However, a plot of microwave spectrum in the absence of rain and other hydrometeors (Figure 1.3) for Kolkata (22°N), India, and Rio de Janeiro, Brazil (23°S), is presented for clarity, taking advantage of almost the same and opposite latitudinal occupancies, on July 1, 2009, during morning hours. Note that at Kolkata and Rio de Janeiro, the weak resonant line is present at 22.234 GHz at the surface. The corresponding peak attenuation at Kolkata is 0.6 dB/km, but at Rio de Janeiro it is 0.3 dB/km. The next maxima are at 60 GHz for both places, and each of them suffers 10 dB/km attenuation. But at around 10 km height from the surface, at Kolkata, a resonant line at 22.234 GHz with 0.005 dB/km attenuation is prominent. On the other hand, the absence of this line is prominent at Rio de Janeiro. The reason behind the absence of the 22.234 GHz resonance line at Rio de Janeiro at 10 km height may be due to the fact that no or very little transportation of water vapor from the surface takes place at that height over Rio de Janeiro at the time in question.

It has been shown by Karmakar et al. (2010) that the integrated water vapor content over Brazil during morning hours is about 40 kg m^{-2}, and that over Kolkata this value goes up to 60 kg m^{-2} (Karmakar et al., 1994). It is discussed by Sen et al. (1989) that any transportation of water vapor from the surface to the higher altitudes has a negligible effect within a 12- to 24-hour timescale. If the timescale is increased

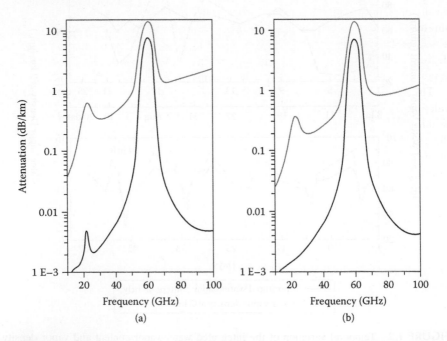

FIGURE 1.3 Microwave spectrum in the absence of rain and other hydrometeors for (a) Kolkata (22°N) and (b) Brazil (23°S).

to 48 hours, the integrated water vapor content is poorly correlated with that around 2 km height, which happens to be the scale height over Brazil (Karmakar et al., 2010). The difference of behavior in this type of variation for a short (12–24 hours) and a long (48 hours) time suggests that the transportation of water vapor to high altitudes occurred within a timescale greater than 24 hours. In other words, we can say that most of the water vapor over Brazil remains within the troposphere. The Clausius-Clapeyron relation establishes that air can hold more water when it warms. This and other basic principles indicate that warming associated with increased concentrations of the other greenhouse gases will also increase the concentration of water vapor. This suggests that in the morning spell the place of choice at Brazil is cooler than that at Kolkata.

Microwave systems are only now starting to be used successfully, yet they have a past that is nearly as long as that of radio waves. For most of the last century, microwaves were not actively exploited, and it is only in the last few years that much research and development has led to viable systems. Microwaves are now set to play an increasing part in radar, communication, and remote sensing (Olver, 1989).

The term *microwave* does not have a very precise definition, but the accepted division of the electromagnetic spectrum is 3–30 GHz. The technology that is being used now offers large application potential in this band. The measurement techniques divide according to the wavelength sensitivity of the detecting apparatus. For example, one of the principal advantages of using microwave systems is their all-weather capability. The principal disadvantage of microwave systems is their sensitivity to propagation path effects that are induced by water vapor. But on the other hand, this becomes a boon to remote sensing scientists as far as measurement of water vapor content or distribution is concerned. Water vapor is not a well-mixed atmospheric constituent, and it is impossible to model it with a high degree of accuracy. But still, the most promising technique available for making the vapor measurement is the use of passive microwave radiometry.

The 22.234 GHz line is suitable mainly for ground-based study, which can provide valuable information regarding the total water vapor content, diurnal variation of water content, and even the vertical distribution, with some simplifying assumptions that the signal-to-noise ratio is largest at this frequency, provided the vertical profiles of pressure and temperature are constant (Resch, 1983). But this does not happen in practice. Westwater (1978), by using Van Vleck line shape along with Liebe's parameter, and again by using Zhevaking-Naumov Gross line shape along with the parameter given by Waters, showed that the frequency, independent of pressure, lie both ways around the 22.234 GHz line. This single-frequency measurement will be influenced by the presence of overhead cloud liquid. Moreover, it has been shown by Simpson et al. (2002) that a zenith-pointing ground-based microwave radiometer measuring sky brightness temperature in the region around 22 GHz is three times more sensitive to water vapor than that to liquid water. But in the region of 30 GHz, the sky brightness temperature is two times more sensitive to liquid water than to water vapor. However, such studies are only not useful for earth-space paths and horizontal links, but also for much moisture-related processes (Vandana, 1980). These types of studies are important for cloud-seeding experiments and for many meteorological research studies (Bhattacharya, 1985).

Out of all the meteorological parameters, water vapor and nonprecipitable cloud liquid water, along with ambient temperature, are found to be the most important for controlling thermodynamic balance, photochemistry of the atmosphere, the sun-weather relationship, and the biosphere. Measurements of the vertical and horizontal distribution of water vapor, as well as its temporal variation, are essential for probing into the mysteries of several atmospheric effects. In this context, ground-based microwave radiometric sensing appears to be a good solution for continuous monitoring of atmospheric water vapor. Radiometric data have been extensively used by quite a few investigators (Westwater et al., 1990, 2005; Gordy, 1976; Gordy et al., 1980, Westwater and Guiraud, 1980; Pandey et al., 1984; Janssen, 1985; Cimini et al., 2007; Karmakar et al., 1999, 2001, 2011(a), 2011(b); Sen et al., 1990) to determine water vapor budget.

1.3   ATMOSPHERIC REMOTE SENSING

Neutral air is comprised mainly of 78% nitrogen, 21% oxygen, and very little argon, carbon dioxide, and water vapor. But water vapor, although such a little amount, controls the thermodynamic balance of the atmosphere. It is well known that water vapor accounts for the largest percentage of the greenhouse effect, between 36 and 66% for water vapor alone, and between 66 and 85% when factoring in clouds. The overall effect of all clouds together is that the earth's surface is cooler than it would be if the atmosphere had no clouds (Karmakar, 2011). Water vapor concentrations fluctuate regionally, but human activity does not significantly affect water vapor concentrations except at a local scale, such as near irrigated fields. Since water vapor is a greenhouse gas and because warm air can hold more water vapor than cooler air, this amplifies the original warming due to water vapor. Another important consideration is that since water vapor is the only greenhouse gas whose concentration is highly variable in space and time in the atmosphere, its real-time measurement is an emerging topic of intense research interest being pursued in different locations all over the globe. The IPCC Fourth Assessment Report (Cracknell and Varotsos, 2007) says that a further warming of about 0.1°C per decade would be expected even if the concentration of all greenhouse gases and aerosols had been kept constant. This report also says that in order to reduce the level of existing uncertainties, the modeling of nature-society interaction is urgently required on a long-term basis, taking into account nonlinear changes in climate systems.

The atmosphere is composed of several layers, namely, the troposphere, stratosphere, mesosphere, and thermosphere, in order of altitude. The tropopause is the thin layer that divides the troposphere and stratosphere. It occurs at about 8 km in the polar region and 16 km in the equatorial region. The temperature of the troposphere decreases by about 0.6°C with every 100 m increase in altitude. Plots of temperature profiles over Kolkata (22°N), Jodhpur (26°N), and Srinagar (34°N) are shown for clarity (Figure 1.4). The temperature in the polar region is low year-round and high near the equator. This temperature difference produces atmospheric circulation. The majority of the meteorological phenomena happen to be due to fluctuation of vapor and temperature as well. Convection is the phenomenon where the air

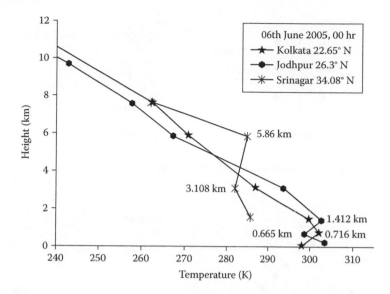

FIGURE 1.4 Temperature profile over Kolkata (22°N), Jodhpur (26°N), and Srinagar (34°N).

at low altitude ascends when it is heated and becomes lighter, and air surrounding it descends to occupy the vacant space. Convection gives rise to ascending currents that form cloud and rain.

In this book we will discuss in detail the sensing methodology of vapor, rain, and cloud and the results obtained in different latitudinal occupancy.

Various methods are available for sensing the atmosphere. The instruments used are called sensors. If the instrument consists of both a transmitter and a receiver, then we call it an active sensor. On the other hand, if the external signal is picked up by a specially designed receiver, we call it a passive sensor (Hartl, 1987). An arbitrary sketch (Figures 1.5 and 1.6) will be helpful in understanding the sensing methodology. The passive sensing may provide us the amount of signal emitted from the suspended atmospheric particles. In our work, we will emphasize the emission from the suspended particles in the atmosphere, namely, water vapor, oxygen, rain, and cloud liquid.

The basic method we will discuss here is radiometry, along with the use of radiosonde data at the place of choice. Specifically, this work concentrates on the recent developments of ground-based microwave radiometry along with some basic terms and principles used in atmospheric remote sensing.

1.4 ATMOSPHERIC INFLUENCES ON THE ELECTROMAGNETIC SPECTRUM

The ever-increasing demand in exploiting radio spectrum at the lower frequency has resulted in increased activity in the millimeter-wave band, particularly for communication systems. Hence, the significant revision of the International Table

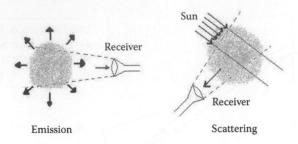

Emission   Scattering

FIGURE 1.5   Pictorial presentation of passive sensing.

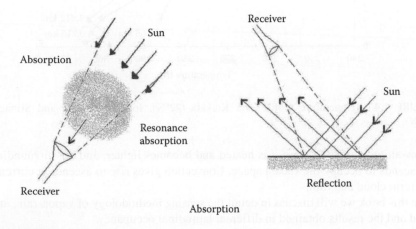

FIGURE 1.6   Pictorial presentation of active sensing.

of frequency allocation above 40 GHz was urgently needed (Katzenstein et al., 1981). But at these frequencies the atmospheric attenuation remains the biggest stumbling block, and hence creates problems of transmission fidelity, reliability, and cost also. However, this may be surmounted by using low cost with short-range repeaters, which are of the order of few kilometers. In fact, short-range systems of 8–12 km with small outage times are possible even in adverse weather, with large ranges being available in good weather (Tsao et al., 1968). In this context, we can look at the millimeter-wave band radio systems (Chang and Yuan, 1980) such as mobile intercept-resistant radio (MISR). Looking at this highly emergent and useful radio spectrum (millimeter-wave band), we need to determine the window frequencies where the atmospheric attenuation is least at the particular place of choice. However, Button and Wiltse (1981) indicate that the attenuation rate (dB/km), also called the specific attenuation, exhibits minima around 30, 94, and 140 GHz. But at a given frequency, the atmospheric absorption coefficient is a function of three basic parameters: temperature, pressure, and water vapor density. The water vapor absorption essentially depends on water vapor density, and likewise, the oxygen absorption depends on pressure and temperature. Since water vapor density and pressure decrease exponentially with increasing altitude, the major contribution in finding the zenith opacity (dB) is provided by the layers closest to the surface (Ulaby et al., 1981). Hence, it is expected that the zenith opacity will vary linearly with the surface water vapor density, which is supported experimentally by Waters (1976). Keeping all this in mind, an attempt has been made to find out the window frequency lying between the two maxima occurring at 60 and 120 GHz, which eventually happened due to oxygen present in the atmosphere. For this purpose, eight places have been chosen, of which four are from the northern latitude and the other four from the southern latitude. The latitudinal occupancy is presented in Table 1.1. In doing so, it was kept in mind that Chongging, China, and Porto Alegre, Brazil, lie at the same latitude, China from the northern latitude and Brazil the southern latitude. The radiosonde data over the above said places were made available by the British Atmospheric Data Centre (BADC). It is to be noted that since the data from corresponding meteorological stations were not available, liberty has been taken to use the surface data from BADC pertaining to the present study, although BADC provides the vertical profiles of the atmospheric data. For clarity, the absorption spectra up to 200 GHz over Kolkata have been presented in Figure 1.7, using the updated millimeter-wave propagation model (MPM) as described by Liebe (1985). In fact, from 1990 onward, International Telecommunication Union (ITU-R) has adopted the MPM model in a somewhat truncated form to describe the absorption behavior of millimeter-wave propagation, in which the surface values of atmospheric parameters like temperature, pressure, and humidity were used as input parameters to determine the specific attenuation profile in the millimeter-wave band.

1.4.1 TEMPERATURE AND HUMIDITY VARIATION OVER A FEW PLACES OF NORTHERN AND SOUTHERN LATITUDES

Radiosonde data, available from BADC (UK), consist of vertical profiles of temperature, $t(h)$, in degrees centigrade, pressure, $P(mb)$, and dew point temperature, t_d, in degrees centigrade over the places of choice. Using these data for the year 2005, the water vapor pressure, $e(mb)$, and saturation water vapor pressure, $e_s(mb)$, using

TABLE 1.1
Latitudinal Occupancy of Eight Places over the Globe

Place	Country	Latitude
Kolkata	India	22.65°N
Chongging	China	29.0°N
Srinagar	India	34.0°N
Aldan	Russia	58.0°N
Lima Callao	Peru	12.0°S
Porto Alegre	Brazil	29.0°S
Paraparaumu	New Zealand	40.0°S
Comodoro Rivadavia	Argentina	45.0°S

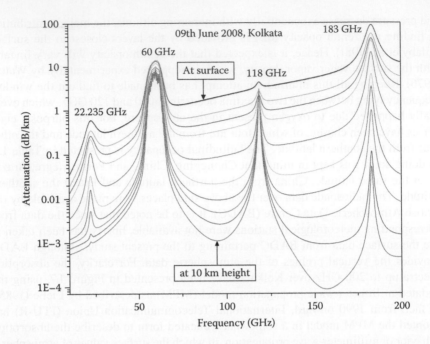

FIGURE 1.7 Microwave absorption spectra over Kolkata at different heights.

the following relations, respectively (Buck, 1981), have been computed by Karmakar et al., 2011:

$$e(mb) = 6.105 \; exp \; \{25.22[1 - 273 \, / \, T_d] - 5.31 \; loge[T_d \, / \, 273]\} \qquad (1.1)$$

$$e_s(mb) = 6.1121 \; exp\{(17.502t) \, / \, (t + 240.97)\} \qquad (1.2)$$

Here, T_d is the dew point temperature in Kelvin, and t is ambient temperature in degrees centigrade. Water vapor pressure, e, and water vapor density, ρ (g/m³), are related by

$$\rho = 217e/T(K) \qquad (1.3)$$

The relative humidity, RH (%), is given by the relation

$$RH = (e \, / \, e_s) \times 100 \qquad (1.4)$$

Figure 1.8 shows the temperature variation over the entire range of latitude as chosen. The interesting point to be noted here is that the temperatures at Kolkata and Brazil are almost equal during the months of January–February. But on the other hand, temperature differs appreciably over these two places during the months

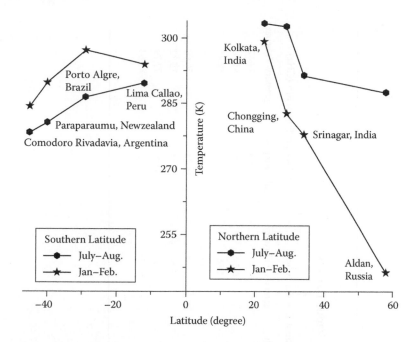

FIGURE 1.8 Variation of temperature with latitude at different locations for the months of January–February and July–August.

July–August, although the latitudinal occupancy is equal (Table 1.2). It is to be noted here that the span of months in a year is so chosen that January–February is considered to be the prevailing winter months over northern latitudes but the rainy season over southern latitudes. The reverse is true for July–August. Again, over the places of exactly the same latitude, i.e., over Chongging, China, and Porto Alegre, Brazil, the temperature difference between the two places is always the same, which is approximately equal to 15 K.

This ensures that although the latitudes are the same, the absolute temperatures depend mostly on the prevailing climatological conditions. Now, taking Kolkata and Brazil into account in our objective, it is seen that at Kolkata the radiative heating of the atmosphere is maximum, as Kolkata is mostly populated with buildings having a dearth of trees and water tanks. But at the same time, if we look at Srinagar, the temperature is also less than that at Kolkata, as Srinagar is surrounded by hills and valleys that are mostly covered with plants and trees. Again, if we look at Brazil, the temperature is still less, as this place is full of small hills covered with vegetation.

Now referring to Figure 1.9, the water vapor density always bears a high value over the southern latitude during January and February, which are considered to be the rainy season there. But on the other hand, the water vapor density is less during those months, as they are considered to be the winter season in the northern latitudes. However, among the places at northern latitudes, Kolkata bears the maximum water vapor density and Russia bears the minimum. The situation is reversed during the months of July–August (Figure 1.9).

TABLE 1.2

Variation of Window Frequency with Temperature and Surface Water Vapor Density

Place	Country	Temperature (K)		Water Vapor Density (g/m³)		Window Frequency (GHz)	
		July–August	January–February	July–August	January–February	July–August	January–February
Kolkata	India	303.45	299.15	24.39	12.91	73.69	77.42
Chongqing	China	302.72	282.81	20.04	6.57	74.48	81.50
Srinagar	India	291.22	277.58	13.71	4.86	75.98	82.46
Aldan	Russia	287.47	246.34	10.13	0.45	78.15	94.26
Lima Callao	Peru	289.60	293.85	10.59	15.54	78.25	76.00
Porto Alegre	Brazil	286.49	297.21	11.02	17.84	78.24	75.32
Paraparaumu	Newzeland	280.47	289.70	6.75	11.75	81.5	77.79
Comodoro Rivadavia	Argentina	278.31	284.29	3.79	6.97	85.35	81.64

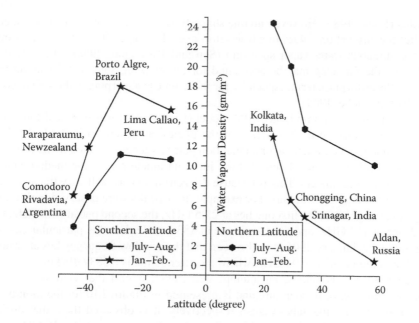

FIGURE 1.9 Variation of water vapor density with latitude at different locations for the months of January–February and July–August.

1.4.2 Determination of Window Frequencies in the Electromagnetic Spectrum

The radiosonde data (surface data) available from BADC for the months of July–August and January–February of the year 2005–2006 have been used in 58°N through 45°S latitudes covering the places as described in Table 1.1. Eventually, it is to be mentioned here, during the months of July–August, the northern latitudes become hot and humid, and the reverse is true for the southern latitudes. On the other hand, during the months of January–February the southern latitudes become hot and humid, and the reverse is true for the northern latitudes.

These particular months were chosen to visualize the climatological conditions over certain parts of the globe and the usable window frequencies between a pair of conventional spectroscopic millimeter-wave absorption maxima.

1.4.2.1 Background Methodology in Determining Window Frequency

A water vapor molecule behaves as an electrical dipole and interacts with the electric field vector of the incoming electromagnetic wave at the microwave and millimeter-wave band, which is subject to the change of quantum level of the water vapor molecule at the same specific frequencies. However, the frequency dependence of absorption is a function of line-width parameter, pressure, temperature, and humidity of the atmosphere (Van Vleck and Weisskopf, 1954). But the Van Vleck–Weisskopf line shape function was modified in order to describe the overlap effects to a first-order approximation. This leads to the local line absorption profiles (Liebe, 1985). However, no line shape function has been confirmed that can predict the absorption values in a true sense over the range 10^{-3} to 10^{-6} as required for a continuous water vapor spectrum (Sen and Karmakar, 1988). The contribution from the far wing may be accounted for by adopting some empirical corrections to the absorption term, especially for those whose absorption values go beyond 105 dB/km (Liebe, 1985).

Keeping all this in mind, the MPM model (Liebe, 1985) was used and found the absorption spectra up to 200 GHz over the places, as described in Table 1.1. The total attenuation, i.e., water vapor attenuation plus oxygen attenuation in dB/km, was calculated. It is to be noted here that the input parameters were the median values of the surface parameters. This is because the zenith opacity will vary linearly with the surface water vapor density. For example, four peaks were observed within this range over Argentina; the first one lies at 22.5 GHz, the second one at 60.5 GHz, the third one at 119 GHz, and the fourth one at 183.5 GHz. For this particular case of Argentina, for January 2006, it is also seen that the window frequency lies at around 80 GHz in the vicinity of 94 GHz, where the attenuation is 0.3839 dB/km.

Using this methodology, an attempt was made to find the window frequencies as a function of temperature and is presented in Figure 1.10 for the months of January–February and July–August, respectively. It is observed there that during January–February, the highest values of window frequencies are 94.26 GHz over Russia (58°N) and 81.64 GHz over Argentina (45°S). Similarly, during July–August, the highest values of windows over the same places are described in Table 1.2. Another attempt was made to work out regression analyses (Figure 1.10) regarding

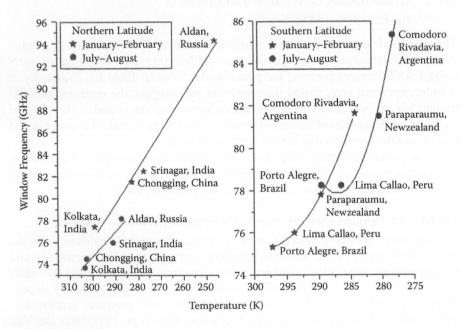

FIGURE 1.10 Variation of window frequency with temperature at different locations for the months of January–February and July–August.

the allocation of window frequency (GHz) with temperature, T (K), yielding the following best-fit relations along with correlation coefficient, R^2:

Northern latitude:

$$f_{window} \text{ (January–February)} = 173.80 - 0.3251 \times T \qquad (R^2 = 0.9925)$$

$$f_{window} \text{ (July–August)} = 144.01 - 0.231 \times T \qquad (R^2 = 0.952)$$

Southern latitude:

$$f_{window} \text{ (January–February)} = 2733.9 - 17.7851 \times T + 0.0297 \times T^2 \quad (R^2 = 1)$$

$$f_{window} \text{ (July–August)} = 7766.09 - 53.55 \times T + 0.0932 \times T^2 \qquad (R^2 = 0.9869)$$

A similar attempt has been made to find out the variation of window frequencies as a function of surface water vapor density and is presented in Figure 1.11. In this case, we find the minimum value of window frequency lies at Kolkata and Brazil, where the water vapor densities are maximum for the months of January–February.

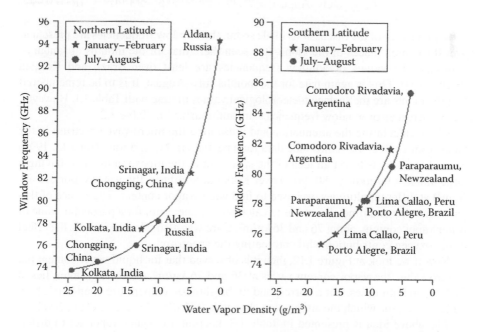

FIGURE 1.11 Variation of window frequency with water vapor density at different locations for the months of January–February and July–August.

On the other hand, where the water vapor density is minimum (see Table 1.2), the value of the window frequency is maximum. Eventually it happens also at Russia.

The situation is similar for the months of July–August. The exception here (Figure 1.11) is that in the southern latitude, the values of the window frequencies are always higher than those over northern latitudes. Table 1.2 comprehensively presents these for clarity.

A similar attempt was made to work out regression analyses (Figure 1.11) of window frequency, f_{window} (GHz), with water vapor density, ρ (g/m³), yielding the following best-fit relations with correlation coefficient:

Northern latitude:

$$f_{window} \text{ (January–February)} = 76.55 + 19.58 \exp(-\rho/4.35) \quad (R^2 = 0.9958)$$

$$f_{window} \text{(July–August)} = 73.21 + 22.54 \exp(-\rho/6.64) \qquad (R^2 = 0.996)$$

Southern latitude:

$$f_{window} \text{ (January–February)} = 73.22 + 20.61 \exp(-\rho/7.78) \quad (R^2 = 0.999)$$

$$f_{window} \text{ (July–August)} = 77.39 + 26.59 \exp(-\rho/3.14) \qquad (R^2 = 0.999)$$

In pursuance of the discussions made so far, the window frequencies are presented over the chosen places of the globe. It is seen there that for the months of January–February, the values of window frequencies are least over Brazil and Kolkata (Figure 1.11). This is again true for the months July–August. It is to be remembered that the inputs are the surface meteorological values in tune with Table 1.1. However, the occurrences of window frequencies are summarized in Table 1.2.

While calculating the attenuation and presenting the microwave spectrum in the band from 1–40 GHz, radiosonde data for a particular day and time (June 15, 1996) were chosen, so as to get an idea of the spectrum in the monsoon months (having a large ambient humidity, ~80–90%) over Kolkata. Similarly, the water vapor density (g/m³) at different altitudes was used to find water vapor content. The choice of this particular day and time allowed us to calculate the spectrum for a particular value of water vapor content, about 76 and 36 kg/m². Care was also taken to include the effect of the oxygen contribution while presenting the spectrum.

Now if we look at Figure 1.12, then it is observed that for liquid water content put at zero value, the vapor content varies at 36 and 76 kg/m² and the minimum lies at 31 GHz in both cases, but a corresponding brightness temperature change of about 10 K occurs, for which the attenuation change is 0.11 dB (Karmakar et al., 2002).

The above idea is presented in Table 1.3, keeping the water vapor set to different values for different days and time. The table shows the variation of minima and peak, although small, with the variation of water vapor content. However, the

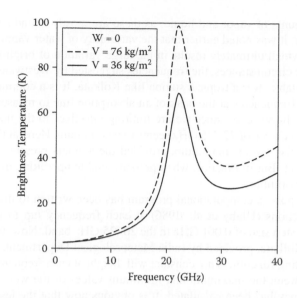

FIGURE 1.12 The brightness temperature as a function of frequency for constant liquid water content and variable values of water vapor content: 36 and 76 kg/m².

TABLE 1.3

The Variation of the Peak and Window Frequencies with Variable Liquid Water Content

	Liquid kg/m²	Vapor kg/m²	Peak Frequencies (GHz)	Window Frequencies (GHz)
A	0	76.0	22.416, 60.375, 118.756, 183.336	31.448, 75.212, 124.578, 211.0
B	0	77.4	22.472, 60.375, 118.755, 183.340	31.475, 74.975, 124.5, 210.88
C	0	24.84	22.611, 60.365, 118.755, 183.345	30.525, 82.115, 128.05, 213.295
D	0	27.77	22.449, 60.365, 118.755, 183.337	30.695, 82.236, 128.065, 213.139

A: Date: July 15, 1991; time: 1700 IST; surface temperature: 303.6 K; surface vapor density: 23.5 g/m³.
B: Date: July 15, 1991; time: 0500 IST; surface temperature: 303 K; surface vapor density: 22.3 g/m³.
C: Date: January 17, 1991; time: 0500 IST; surface temperature: 287 K; surface vapor density: 10.81 g/m³.
D: Date: January 17, 1991; time: 1700 IST; surface temperature: 293.7 K; surface vapor density: 7.61 g/m³.

---

constancy of weighting function [K (g/m³) km⁻¹] with height at the desired frequency provides the potential for application of that frequency. Westwater et al. (1990) showed that below about 5 km, the water vapor and liquid water weighting function at 20.6 and 31.65 GHz are nearly constant with height. This shows that the variations in brightness temperature are primarily affected by variations in the column integrated amounts of vapor and liquid, as these are mostly abundant below about 5 km. However, it was noted earlier that the variability of water vapor over Kolkata is very large, which ultimately results in the large variation of brightness temperature. Under the circumstances, the frequencies suggested by Westwater et al. (1990) may not be suitable over a tropical station like Kolkata. It is a common practice to use one of the frequencies at the peak of an absorption line to measure the density of a substance. In our case, since we are limiting ourselves to the measurement of water vapor, the choice of 22.235 GHz seems to be optimum. Here, at this frequency, the signal-to-noise ratio is maximum, provided the ambient pressure and temperature are constant. But in practice, when pressure and temperature are variable, the selection is not optimum.

For this purpose, a computational program has been written to find water vapor weighting functions (Ulaby et al., 1986) at each frequency (up to three decimal places), with a step size of 0.001 GHz in the 20–25 GHz band. Now, from the upper air data over Kolkata, provided by India Meteorological Department, it is observed that the weighting function is not constant with height at each frequency. Hence, the difference between the maximum and minimum values of the weighting function at each frequency has been calculated. It is obvious now that the lesser the difference, the steeper is the weighting function, and vice versa. Keeping this idea in mind, plots of frequency vs. this difference have been prepared (Figure 1.13) for the months of January and July 1996, as these two months are found to be the maximum and minimum, respectively, bearing vapor over Kolkata. In pursuance of the previous discussions, the window frequencies are presented over the chosen places of the globe. It is seen there that for the months of January–February, the values of window frequencies are least over Brazil and Kolkata. This is again true for the months July–August.

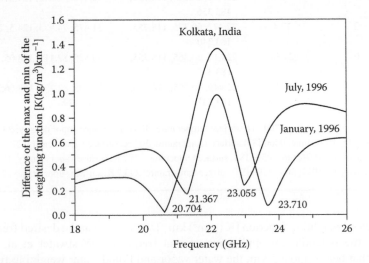

FIGURE 1.13 The difference between the maximum and minimum values of the water vapor weighting function [K(g/m^3)km^{-1}] as a function of frequency.

Comparing Figures 1.10 and 1.11, it is seen that variation of window frequencies in lieu of conventional 94 GHz over different places in the globe is mostly influenced by water vapor density rather than temperature.

To clarify this, let us take an example of Srinagar, India, and Porto Alegre, Brazil. At Srinagar the fractional changes in temperature and water vapor density (July through January) are 0.74 and 0.64, respectively, and the corresponding fractional shift in window frequency is 0.078. So it is seen that the fractional change in water vapor density is lower than that of temperature for the same shift in window frequency.

Similarly, over Brazil the fractional changes in temperature and water vapor densities are 0.44 and 0.38, respectively, and the corresponding shift in window frequency is 0.037. It is to be noted that although the latitudinal occupancies of Srinagar and Porto Alegre are almost the same, the variability of water vapor density is a little higher at Srinagar than that at Porto Alegre, which influences the shifting of conventional window frequency. It is also clear that the months January–February are winter for the northern hemisphere and summer for the southern hemisphere. And the months July–August are the opposite. It does not matter if two locations are in the same latitude but in different hemisphere for the same period. It is also to be mentioned that the meteorological equator is not located exactly over the equator, and it moves toward the northern hemisphere during July–August and toward the southern hemisphere during January–February. However, the most relevant fact for water vapor distribution is the circulation pattern (i.e., wet or dry season) and the season (winter or summer).

It has been assumed that the dimensions of water drops are much less than the wavelength of the incoming signal. Here, in this context, it has been assumed that liquid water exists only when the relative humidity exceeds a certain value. It has also been assumed that the lower limit for relative humidity within a typical type of cloud is 95%, with water particles having a diameter range of 10–50 μm. However, for remote sensing purposes, it is always suggested that the window frequency suitable for a particular place of choice be selected, which in turn depends on the liquid water content over a particular location.

REFERENCES

Askne, J.I.H., and E.R. Westwater. A review of ground based remote sensing of temperature and moisture by passive microwave radiometers. *IEEE Trans. Geosci. Remote Sensing*, GE-24(3), 340–352, 1986.

Bhattacharya, C.K. Radiometric studies of atmospheric water vapour and attenuation measurements at microwave frequencies. PhD thesis, Banaras Hindu University, India, 1985.

Buck, A.L. New equations for computing vapour pressure and enhancement factor. *J. Appl. Meteorol.*, 20, 1527–1532, 1981.

Button, K.J., and J.C. Wiltse. *Infrared and millimetre waves*. Vol. 4. Academic Press, New York, 1981.

Chang, Y.W., and L.T. Yuan. Millimeter wave binocular radio. *Microwave J.*, 3, 31–36, 1980.

Cimini, D., E.R. Westwater, A.J. Gasiewski, M. Klein, V. Leuski, and J.C. Liljegren. Ground-based millimeter and submillimeter-wave observations of low vapour and liquid water contents. *IEEE Trans. Geosci. Remote Sensing*, 45, 2169–2180, 2007.

Cracknell, A.P., and C.A. Varotsos. The IPCC Fourth Assessment Report and the fiftieth anniversary of Sputnik. *Environ. Sci. Pollut. Res.*, 14, 384–387, 2007.

Emery, R.J., and A.M. Zavody. Atmospheric propagation in the frequency range 100–1000 GHz. *Radio Electronic Eng.*, 49(7), 370–380, 1979.

Gordy, C. Remote sensing of the atmospheric water content from satellite using microwave radiometry. *IEEE Trans. Antennas Propagation*, 24, 155–162, 1976.

Gordy, N.C., A. Gruber, and W.C. Shen. Atmospheric water content over the tropical Pacific derived from Nimbus-6 scanning microwave spectrometer. *J. Appl. Meteorol.*, 19(8), 968–996, 1980.

Hart, L. Ph. Fundamentals of remote sensing. In *Proceedings of a course "Application of Remote Sensing to Agrometeorology,"* ed. F. Toselli. Kluwer Academic, Dordrecht, The Netherlands, 1987.

Janssen, M.A. A new instrument for the determination of radio path delay due to atmospheric water vapour. *IEEE Trans. Geosci. Remote Sensing*, 23, 455–490, 1985.

Karmakar, P.K. *Microwave propagation and remote sensing: Atmospheric influences with models and applications.* CRC Press, Boca Raton, FL, 2011.

Karmakar, P.K., S. Chattopadhyay, and A.K. Sen. 1999. Estimates of water vapour absorption over Calcutta at 22.235 GHz. *Int. J. Remote Sensing*, 20, 2637–2651, 1999.

Karmakar, P.K., S. Devbarman, S. Chattopadhaya, and A.K. Sen. Radiometric studies of transportation of water vapour over Calcutta. *Int. J. Remote Sensing (UK)*, 15(7), 1637–1641, 1994.

Karmakar, P.K., M. Maity, A.J.P. Cahiers, C.F. Angelis, L.A.T. Machado, and S.S. Da Costa. Ground based single frequency microwave radiometric measurement of water vapour. *Int. J. Remote Sensing (UK)*, 32(23), 1–11, 2011a.

Karmakar, P.K., M. Maity, S. Chattopadhyay, and M. Rahaman. Effect of water vapour and liquid water on microwave absorption spectra and its application. *Radio Sci. Bull.*, 303, 2002.

Karmakar, P.K., M. Maity, S. Mondal, and C.F. Angelis. Determination of window frequency in the millimeter wave band in the range of 58 degree north through 45 degree south. *Adv. Space Res.*, 48, 146–151, 2011b.

Karmakar, P.K., M. Maiti, S. Sett, C.F. Angelis, and L.A.T. Machado. Radiometric estimation of water vapour content over Brazil. *Adv. Space Res.*, 48, 1506–1514, 2011c.

Karmakar, P.K., M. Rahaman, and A.K. Sen. Measurement of atmospheric water vapour content over a tropical location by dual frequency microwave radiometry. *Int. J. Remote Sensing*, 22(17), 3309–3322, 2001.

Karmakar, P.K., L. Sengupta, M. Maiti, and C.F. Angelis. Some of the atmospheric influences on microwave propagation through atmosphere. *Am. J. Sci. Ind. Res.*, 1(2), 350–358, 2010.

Katzenstein, W.E., R.K. Moore, and H.G. Kimball. Spectrum allocations above 40 GHz. *IEE Trans. Commun.*, 29, 1136–1141, 1981.

Liebe, H.J. An updated model for millimeter wave propagation in moist air. *Radio Sci.*, 20, 1069–1089, 1985.

Olver, A.D. Millimetre wave systems—Past, present and future. *IEE Proc.*, 136(Pt. F, 1), 1989.

Pandey Prem, C., B.S. Gohil, and T.A. Hariharan. A two frequency logarithmic differential technique for retrieving precipitable water from satellite microwave radiometer (SAMIR-II) on board Bhaskar II. *IEEE Trans. Geosci. Remote Sensing*, 22, 647–655, 1984.

Resch, G.M. *Another look at the optimum frequencies for water vapour radiometer.* TDA Progress Report. TDA, New York, 1983.

Rogers, T.F. Calculated centimetre-millimeter water vapour absorption at ground level. Presented at Proceedings of the Conference on Radio Meteorology, University of Texas, Texas, 1953.

Rogers, T.F. Absolute intensity of water vapour absorption at microwave frequencies. *Phys. Rev.*, 93, 248–249, 1954.

Sen, A.K., and P.K. Karmakar. Microwave communication parameters estimated from radiosonde observations over the Indian sub-continent. *Ind. J. Radio Space Phys.*, 17, 1988.

Sen, A.K., P.K. Karmakar, T.K. Das, A.K. Devgupta, P.K. Chakraborty, and S. Devbarman. Significant heights for water vapour content in the atmosphere. *Int. J. Remote Sensing*, 10, 1119–1124, 1989.

Sen, A.K., P.K. Karmakar, A. Mitra, A.K. Devgupta, M.K. Dasgupta, O.P.N. Calla, and S.S. Rana. Radiometric studies of clear air attenuation and atmospheric water vapour at 22.235 GHz over Calcutta. *Atmos. Environ.*, 24A(7), 1909–1913, 1990.

Simpson, P.M., E.C. Brand, and C.L. Wrench. *Liquid water path length algorithm development and accuracy: Microwave radiometer measurements at Chilbolton, Radio Communication Research Unit.* CLRC—Rutherford Appleton Lab, Chilton UK, 2002.

Solheim, F., J.R. Godwin, E.R. Westwater, Y. Han, S.J. Keihm, K. Marsh, and R. Ware. Radiometric profiling of temperature, water vapor and cloud liquid water using various inversion methods. *Radio Sci.*, 33, 393–404, 1998.

Straiton, A.W., and C.W. Tolbert. Anomalies in the absorption of radio waves by atmospheric gases. *Proc. IRE*, 48, 898–903, 1960.

Tsao, C.K.H., J.J. de Battencourt, and P.A. Kullstan. Design of millimeter wave communication system. *Microwave J.*, 11, 47–51, 1968.

Ulaby, F.T., R.K. Moore, and A.K. Fung. *Microwave remote sensing.* Vol. 1. Addison-Wesley, Reading, MA, 1981.

Ulaby, F.T., R.K. Moore, and A.K. Fung. *Microwave remote sensing: Active and passive.* Artech House, Norwood, MA, 1986.

Van Vleck, J.H. The absorption of microwaves by oxygen. *Phys. Rev.*, 71, 413–424, 1947a.

Van Vleck, J.H. The absorption of microwaves by uncondensed water vapour. *Phys. Rev.*, 717 425–433, 1947b.

Van Vleck, J.H., and W.F. Weisskopf. On the shape of collision-broadened lines. *Rev. Modern Phys.*, 17, 227–241, 1954.

Vandana, D. Microwave radiometric studies of atmospheric water vapour and attenuation measurements at 22.235 GHz. PhD thesis, University of Delhi, India, 1980.

Viktorova, A.A., and S.A. Zhevakin. Rotational spectrum of water vapour dimmers. *Radiophysics Quantum Electronics*, 18, 1976.

Waters, J.W. Absorption and emission of microwave radiation by atmospheric gases, In *Methods of experimental physics*, ed. M.L. Meeks, 2–3. Vol. 12, part 2. Academic Press, New York, 1976.

Westwater, E.R. The accuracy of water vapour and cloud liquid determination by dual frequency ground based microwave radiometry. *Radio Sci.*, 13, 677–685, 1978.

Westwater, E.R., S. Crewell, C. Mätzler, and D. Cimini. Principles of surface-based microwave and millimeter wave radiometric remote sensing of the troposphere. *Quaderni Della Società Italiana Di Elettromagnetismo*, 1, 3, 2005.

Westwater, E.R., and Guiraud, F.O. Ground based microwave radiometric retrieval of precipitable water vapour in the presence of cloud with high liquid content. *Radio Sci.*, 13(5), 947–957, 1980.

Westwater, E.R., J.B. Sinder, and M.J. Fall. Ground based radiometric observation of atmospheric emission and attenuation at 20.6, 31.65, 90.0 GHz: A comparison of measurements and theory. *IEEE Trans. Antenna Propagation*, AP-38(10), 1569–1580, 1990.

2 Radiometry

2.1 INTRODUCTION

It is well accepted that electromagnetic energy may be absorbed or reflected, or both simultaneously, upon its incidence on a material body. When the material body is in thermodynamic equilibrium with its environment, the material absorbs and radiates energy at the same rate. If the material reflects nothing and instead absorbs all of the incident radiation, then the material is called a black body. The black body radiation is well explained by Max Planck on the basis of quantum theory. But in practice, all the bodies are not really black. Accordingly, the real body spectrum at the same black body temperature may be compared with the black body spectrum. The real body radiation spectrum depends on polarization, angular variation of the emitted radiation, absorption, and scattering, which in turn are governed by the geometrical configuration of the surface and interior of the medium concerned, and also by the spatial distribution of its dielectric properties and temperature.

Radiometry is the measurement of incoherent electromagnetic radiation from an object obeying the laws of radiation fundamentals. A microwave radiometer is capable of measuring very low-level microwave radiation from an object. Sometimes it is possible to establish a useful relation between the magnitude of the radiation intensity and a specific terrestrial or atmospheric parameter of interest. Once the relation is established, the desired parameter can thus be obtained from microwave radiometer measurements (Shih and Chu, 2002). Applications of the ground-based microwave radiometer to measure meteorological parameters have been widely accepted for years (Janssen, 1993, and references therein). Since the pioneering work of Westwater (1965), atmospheric temperature profiles were retrieved successfully by many scientists from radiometric measurements of gaseous emission using and applying various inversion techniques (Snider, 1972; Westwater et al., 1975; Decker et al., 1978; Askne and Westwater, 1982). The precipitable water vapor and cloud liquid can also be retrieved from the observations of a microwave radiometer operated at selected frequencies (Guiraud et al., 1979; Snider et al., 1980; Westwater and Guiraud, 1980). The retrieved accuracy of precipitable water vapor measured by the microwave radiometer is believed to be the same as or even better than those estimated by radiosonde. However, the increasing demand of multifrequency radiometers in ground-based meteorological measurement has raised the question of their potential use for the retrieval of rainfall parameters (Sheppard, 1996; Marzano et al., 1999, 2002, 2005; Liu et al., 2001; Czekala et al., 2001).

The potential use of the ground-based microwave radiometer may be considered for atmospheric modeling purposes. But from the modeling point of view, the approach to the ambient water vapor, temperature, etc., and rainfall signature requires some thorough insight into the electromagnetic interaction between microwave radiation and the medium concerned, since radiometric response depends on various radiative sources, while radio waves propagate through the atmosphere. But the accuracy of measurement depends on the atmospheric inhomogenity due to the presence of hydro- meteors in different phases. However, for continuous monitoring of the atmosphere for various advantageous reasons, e.g., short-term weather forecasting, air pollution control, and long-term prediction, several studies have been made by using remote- sensing instruments such as radar, lidar, radiometer, etc., by taking into account that they have certain uncertainties. Ware et al. (2003) compared radiometric profiles with radiosonde and forecast sounding in the evaluation of accuracy of radiometric temperature and water vapor, and from this, a case study has been described that showed the improvement from the forecasting on the basis of variational assimilation of radiometric soundings. Chan et al. (2006) presented the performance and applica- tion of an instability index derived from the radiometric study, which is basically the now-casting of heavy rainfall and thunderstorms in Hong Kong. Barbaliscia et al. (1998) retrieved the integrated water vapor and liquid water content by deploying radiometric measurements in Italy. Doran et al. (2002) examined the differences in cloud liquid water path at coastal and inland locations on the north slopes of Alaska using the ground-based radiometer. Yang et al. (2006) also estimated criterion for determining water vapor sources in summer in the northern plateau of Tibet.

2.2 RADIATION FUNDAMENTALS

A physical body at a temperature above the absolute zero temperature possesses thermal energy that is radiated from the body. The observed dependence of this energy on frequency was explained by Planck, who derived the energy bandwidth per unit volume, U, within a cavity enclosed by a body at temperature T to be

$$U(f,T) = \frac{8hf^3}{c^3} \frac{1}{\exp\left(\dfrac{hf}{KT}\right) - 1} \qquad (2.1)$$

where f is the frequency, h is Planck's constant, c is the velocity of light, and K is the Boltzmann constant. If the body is black and not enclosed, but placed in a vacuum, the black body radiation will be emitted in all directions around the body and can be received as emission noise by a sensitive receiving system for the radiation. For the reception of the radiation with an extended aperture such as an antenna, the radiative properties of such a thermal source are best described by an equation for radiated power P per unit area, per unit solid angle, per unit bandwidth, which allows the intercepted power by the sensing aperture to be calculated as

$$P(f.T) = \frac{2hf^3}{c^2} \frac{1}{\exp\left(\dfrac{hf}{KT}\right) - 1} \qquad (2.2)$$

At microwaves and millimeter waves, $hf \ll KT$, so that the Planck expression, Equation 2.2, can be approximated by the simple Rayleigh Jeans expression as

$$P(f,T)=\frac{2KTf^2}{c^2} \qquad (2.3)$$

Equation 2.3 shows the relationship between the emission noise and absolute temperature for a black body at microwaves and millimeter waves. In fact, all bodies are not black; for example, good reflectors, such as metal targets, are poor emitters characterized by an emission factor $\in$ that is less than unity. The radiated power from such objects will be less and can be derived from Equations 2.2 and 2.3 by replacing T by $\in T$. The emissivity $\in$, is unity for a black body and less than unity for all other bodies.

The sun behaves as a black body at a temperature of about 6000 K at C-band and corresponding to the photosphere, and 12,000 K at ku-band frequency corresponding to the chromosphere. The temperature T is often called the brightness temperature of a black body.

For an antenna placed inside a black body enclosure at a temperature T, and terminated by a resistance equal to the radiation resistance of the antenna (matched termination), the black body radiation received by the antenna will heat up the matched termination, which in turn will radiate through the antenna back into the enclosure. Under thermodynamic equilibrium the power received by the antenna from the black body will be equal to that reradiated by the antenna, and the temperature inside the enclosure will be uniform. The radiated power due to the matched termination will then correspond to the temperature T of the black body, and it is called the antenna temperature, which can also be defined as the temperature to which the radiation resistance of the antenna is to be raised to obtain the same noise power from it as that actually received by the antenna. If the temperature distribution of the black body is not uniform but depends on θ and φ, then it can be shown that the antenna temperature, Ta, is given by

$$T_a=\frac{1}{4\pi}\int T(\theta,\varphi)G(\theta,\varphi)d\Omega \qquad (2.4)$$

where $G(\theta,\varphi)$ is the antenna gain in the direction of θ and φ. In deriving this equation, the reception of only one polarized component parallel to the direction of antenna polarization was considered. As the polarization of the thermal source is random, only half the power given by Equation 2.3 is in fact received by a linearly polarized antenna. If the temperature distribution is a slowly varying function of θ and φ, then $T(\theta,\varphi)$ can be taken outside the integral in Equation 2.4 and we get

$$T_a=T(\theta,\varphi)\frac{1}{4\pi}\int G(\theta,\varphi)d\Omega$$

However, as $\frac{1}{4\pi}\int G(\theta,\varphi)d\Omega=1$, from the definition of antenna gain function we have

$$T_a=T(\theta,\varphi) \qquad (2.5)$$

Equation 2.5 is very useful in radio astronomical investigation of the noise temperature due to the galaxy. For this case, the antenna beam employed is in general much sharper than the brightness temperature distribution over the sky due to galactic emission. It may be mentioned here that the galactic emission is not really thermal in origin but is believed to originate from a synchrotron process occurring in the presence of the galactic magnetic field. Nevertheless, the brightness temperature as indicated in Equation 2.5 gives an equivalent black body temperature distribution of the galaxy and is very useful in radiometric measurements of the galactic radio noise at microwaves and millimeter waves.

If the antenna beam width is greater than the angular extent of the source of radiation, the beam is not filled up by the source, and the antenna temperature as given by Equation 2.4 will be reduced by a factor called the beam fill factor due to the presence of the function $T(\theta,\varphi)$. For the detection of solar radio emission as well as for the detection of targets on the earth's surface by a radiometer antenna, the beam-filling factor is often a very useful design parameter. In a radiometer the antenna is connected to a very sensitive super-heterodyne receiver. If the random fluctuation of the d.c.-detected output power of the receiver can be smoothed by a time constant circuit having an integration time of τ seconds, then it can be shown that the minimum detectable change in the antenna temperature is given by

$$T_a = \frac{T_R + T_a}{(B\tau)^{\frac{1}{2}}} \qquad (2.6)$$

where T_R is the receiver noise temperature and B is the radio frequency (r.f.) bandwidth of the receiver. In deriving this equation it was assumed that the receiver pass band is of an ideal square shape and that the integration is also ideal within the shape limit 0 to τ. In practice, neither the r.f. pass band shape nor the integration is ideal, and hence the minimum detectable antenna temperature changes. Then Equation 2.6 will be changed by a factor depending on the shape of the receiver pass band, and in that case, the Equation 2.6 becomes

$$T_a = K_s \frac{T_R + T_a}{(B\tau)^{\frac{1}{2}}} \qquad (2.7)$$

where K_s is the shape factor. Usually $T_a \ll T_R$, and considering the shape factor is equal to unity, Equation 2.7 becomes

$$T_a = \frac{T_R}{(B\tau)^{\frac{1}{2}}} \qquad (2.8)$$

This is the equation for the minimum detectable change in antenna temperature, also called the radiometer sensitivity in a total power radiometer. The time constant τ is typically about 0.1–1 second, with lower values suitable for missile guidance and target seekers. Higher values are suitable for radio astronomical and remote-sensing studies.

2.3 BASIC PARAMETERS OF RADIOMETRIC SENSING

Depending on surface irregularities, the reflected component will behave as a function of the microstructure, the chemical nature, and the biological state of the target. To illustrate, let us take an example of a spongy leaf of a plant with many distributed holes and drops of water along with some pigments. As discussed in the foregoing section, some part of the energy will be absorbed. The leaf would behave as a lossy material depending on frequency of the microwave band, and hence would be a dispersive multireflection phenomenon along with absorption. Here, we will discuss some of the technical terms about radiometric remote sensing.

2.3.1 BRIGHTNESS TEMPERATURE

A radiometer is simply a passive receiver. In the case of a radiometer, the source is the target itself. The energy received by this kind of receiver is due to radiation self-emitted or reflected by the target. The emitted energy, which is attributed to the thermal motion of electrons within the material, is uniformly radiated in all directions and polarizations. The long wavelength approximation of Planck's law gives an expression called Rayleigh-Jeans law:

$$B_{bb} = \frac{2kT}{\lambda^2} W.m^{-2}.Hz^{-1}.sr^{-1} \qquad (2.9)$$

Here, B_{bb} is the black body brightness, K is Boltzmann's constant, and λ is the wavelength in meters. The percentage error in using Equation 2.9 is less than 1% if $\lambda T \geq 0.77.m.K$. For simplicity, if we consider any target in the earth at $300K$, then its corresponding frequency is $f \leq 117\,GHz$. So it covers the entire microwave spectrum, which is usually used in passive radiometric sensing.

It is known that for a given value of physical temperature T of a black body, the maximum power that an object can emit is equal to P_{bb}. This power would be received by a lossless antenna placed inside a chamber whose walls are made of a perfectly absorbing material. In that case, the power received is

$$P_{bb} = kTB \text{ watt} \qquad (2.10)$$

Here B, is the receiver bandwidth. This expression (2.10) corresponds to power and temperature, which has led to the definition of radiometric temperature. Furthermore, since B is a receiver rather than a target parameter, the use of T rather than P_{bb} to describe the emission from a material surface is more desirable (Ulaby et al., 1986).

In reality, we mostly encounter the real targets instead of a black body. To accommodate real targets, the temperature T is replaced by an equivalent brightness temperature T_b, which would produce the same power P_{bb} as mentioned in Equation 2.10. It is to be mentioned here that we have confined ourselves in the microwave approximation, i.e., tending toward the Rayleigh-Jeans limit. The significant feature of this limit is the linear relationship of Planck's function with physical temperature, which in turn suggests a scaling of intensity I_f as

$$T_b = \frac{\lambda^2}{2k} I_f \tag{2.11}$$

Henceforth, we will use the definition of brightness temperature, which is normally termed the Rayleigh-Jeans brightness temperature.

However, the definition of brightness temperature is not unique (Janssen, 1993). Another definition of brightness temperature is given by the Institute of Electrical and Electronics Engineers as the temperature of a black body radiator that produces the same intensity as the source under observation. The relationship between the two definitions can be expressed (Janssen, 1993) for the emission of a black body radiator at temperature T as

$$R(f,T) = \frac{T_b^1 \, (thermodynamic)}{T_b \, (R-J \, equivalent)} \tag{2.12}$$

Now if we look at the fractional difference, we will see $T - T_b / T_b = R - I$ is small either for low frequencies or for high temperature. The difference $T - T_b$ approaches the equivalent frequency difference of 0.024 Hz.

2.3.2 EMISSIVITY

The ratio $T_b / T = e$, is called the emissivity of the material. In general, T_b and e may be the functions of direction and polarization. Since the maximum power can be radiated by a black body, a good approximation to ideal emitters at microwave frequencies is the highly absorbing materials. On the other hand, a highly conductive metal plate has an emissivity of zero, a perfect reflector.

The brightness temperature of the ground surface is a function of ground thermometric temperature and emissivity. But the emissivity, in turn, is a function of ground geometry and electromagnetic properties. The value of T_b can be calculated from the solution of the radiative transfer equation subject to emission, scattering, and certain boundary conditions.

2.3.3 APPARENT TEMPERATURE

This temperature is related to the energy incident upon the antenna. If a source is observed by an antenna of a receiver, then it receives:

1. Self-emitted radiation from the source
2. Upward/downward atmospheric emission
3. Scattered radiation

Now, the sum effect of these radiations will be attenuated as they propagate through the intervening atmosphere between the source and the antenna. If the atmosphere is lossless, then $T_b \cong T_{AP}$, where T_{AP} is the apparent temperature. This condition may be achieved if we are in the limit 1–10 Hz.

2.3.4 Antenna Temperature

We consider the measured output voltage obtained through the receiver as a function of physical temperature of matched load put in place of the antenna. The noise power P_n delivered by the load is proportional to the physical temperature. Now, corresponding to the power P, provided by the antenna to the receiver, a resistor equivalent temperature T_a can be defined such that the noise power delivered by the resistor at that temperature is equal to P. Thus,

$$T_a = \frac{P}{kB} \qquad (2.13)$$

If the scene observed by the antenna beam is characterized by uniform brightness temperature, then $T_a = T_b$. However, in general, T_a represents all radiations incident upon the antenna, integrated over all possible directions and weighted according to the antenna pattern along with effects of the atmosphere (Ulaby et al., 1981).

2.4 GENERAL PHYSICAL PRINCIPLE

The scalar form of the radiative transfer equation is remarkably simple in the Rayleigh-Jeans limit and is considered to be sufficient for the large majority of microwave applications. For a detailed description, see Chandrasekhar (1960) and Ulaby et al. (1981).

The radiative intensity down-welling from the atmosphere and expressed in an equivalent brightness temperature T_b can be written as (Askne and Westwater, 1986)

$$T_b = T_{bg} \exp\left\{-\int_0^\infty \alpha(h')\,dh'\right\} + \int_0^\infty T(h)\alpha(h)\exp\left\{-\int_0^h \alpha(h')\,dh'\right\}dh \qquad (2.14)$$

Here, T_{bg} is the cosmic background radiation. The attenuation coefficient α is a function of different meteorological parameters along with rain.

The radiation is a nonlinear function of the required quantities, and we linearize the expression around a suitably chosen first guess, such as a climatological mean. We describe changes in the brightness temperature around the first guess by means of a weighting function, which expresses the sensitivity of T_b to the variation of the humidity $\Delta\rho(h)$ or the temperature $\Delta T(h)$ around their initial values:

$$\Delta T_b = \int_0^\infty [W_\rho(h,f,\theta)\delta\rho(h) + W_T(h,f,\theta)\delta T(h) + W_L(h,f,\theta)\delta L(h)$$

$$+ W_P(h,f,\theta)\delta P(h)]dh \qquad (2.15)$$

for a certain frequency f and elevation angle θ. Here, ρ, T, L, P stands for water vapor, T for temperature, L for liquid water, and P for ambient atmospheric pressure. However, the weighting functions analyses for humidity $W\rho$ and temperature W_T by Westwater et al. (1990) in Denver, Colorado, showed that below 5 km water vapor and liquid water weighting functions at 20.6 and 31.65 GHz are nearly constant with height. This implies that variations in T_b are primarily affected by variation in the integrated amount of vapor and liquid water. According to them, the weighing functions are given by

$$W_T(h) = \alpha(h)e^{-\tau(0,h)} + e^{-\tau(0,h)}\frac{\partial\alpha(h)}{\partial T}[T(h)$$

$$-T_{bg}e^{-\tau(h,\infty)} - \int_h^{\infty} T(h')\alpha(h')e^{-\tau(h,h)}\,dh']$$

(2.16)

Now, calling the other variables except temperature the generic variables, we write the weighting functions as

$$W_x = e^{-\tau(0,h)}\frac{\partial\alpha(h)}{\partial x}\left[T(h) - T_{bg}e^{-\tau(s,\infty)} - \int_h^{\infty} T(h')\alpha(h')e^{-\tau(h,h)}\,dh'\right] \quad (2.17)$$

To explain more clearly about the weighting function, we take first the tempera-ture weighting function (km^{-1}). If we have a $\delta T(K)$ change in T over a height interval δh(km), the brightness temperature response $\delta T_b(K)$ to this change is $\bar{W}_T \delta T \delta h$ where $\bar{W}_T$ is called the height average of W_T over the height interval δh. For water vapor, similarly, $\delta T_b(K) = \bar{W}_\rho \delta_\rho \delta h$. If the units of ρ are $g.m^{-3}$ and h in km, then their product $\delta V = \delta\rho \cdot \delta h$ has the same unit of mm. Thus, $\delta V T_b(K) = W\rho\delta V$. The weighting func-tions are determined from the height profile of attenuation coefficients at different frequencies. These in turn depend on the water vapor content at the place in question. The constancy of the weighting function with height at a desired frequency provides the application potential of that frequency.

Equation 2.14 and its Rayleigh-Jeans approximation are well discussed by Goody and Yung (1995), and its more general form, including scattering, is discussed by Janssen (1993). The scattering problem is to be taken care of due to ice or melting liquid, depending on the size distribution of the particles.

Information on meteorological variables may be obtained from measurements of radiometric brightness temperature T_b as a function of f and/or θ. Equation 2.14 is used (1) in forward model studies in which the relevant meteorological variables are obtained by radiosonde sounding, (2) in inverse problem and parameter retrieval applications in which meteorological information is inferred from measurements from radiometric brightness temperature T_b, and (3) in system-modeling studies in determining the effects of instrument noise on retrieval and optimum measurement ordinates such as f and/or θ (Westwater et al., 2005). The region around 22.234 GHz is used for water vapor sensing, while the transmission windows near 30–50, 70–100, and 130–150 GHz are used for cloud sensing. The strong absorption line at or near 60 and 118 GHz is used for temperature sensing. Another strong absorption line at 183 GHz is used for a low amount of vapor sensing, such as vapor at high altitudes.

2.4.1 MICROWAVE ABSORPTION AND EMISSION

In the microwave band, the principal sources of absorption and emission are water vapor, oxygen, and cloud liquid. Rain is perhaps the worst offender in the microwave band. Water vapor absorption arises due to weak electric dipole rotational transition, which produces a peak resonance line at 22.234 GHz, along with a much stronger line at 183.311 GHz. The far-wing contribution, extending up to infrared, also plays a major role in producing the continuum of water vapor. Oxygen absorption takes place due to a series of magnetic transitions centered around 60 GHz and the isolated line at 118.75 GHz. There are also resonances by ozone that are important for stratospheric sounding (Gasiewski, 1993).

2.4.1.1 Gaseous Absorption Models

We have discussed so far that water vapor and oxygen are the main source of absorption of microwaves. Detailed calculations of absorption by water vapor and oxygen were done initially by Van Vleck (1947b). The quantum mechanical basis of these calculations, along with the laboratory measurements, led to increasingly accurate calculations of absorptions (Van Vleck and Weisskopf, 1947). Starting with laboratory measurements in the late 1960s and continuing for several years, Liebe (1985b) proposed a model that is extensively used all over the world. One version of the model (Rosenkranz, 1993) was also tested and is used by several scientific communities in remote sensing areas. More recently, Rosenkranz (1998, 1999) developed an improved version of the propagation model that is also extensively used. But the main issue while calculating the absorption is the choice of pressure-broadened line width parameter, which arises due to self-broadening (H_2O-H_2O collisions) and foreign broadening (H_2O-other molecule collisions). Rosenkranz based his model on the parameters used by Liebe and Layton (1987).

Another model (Liebe et al., 1991; Leuskii et al., 2000) is also used, especially in the U.S. climate research community (Westwater et al., 2005). Recently, two refinements of absorption models have taken place. The first one was done by Clough et al. (2005) and Tretyakov et al. (2005), which is, in fact, refinement of Rosenkranz's 1998 code. Another was done by Liljegren et al. (2005); it incorporated the line width parameters of the 22.235 GHz model with different continuum formalism (Westwater et al., 2005).

2.4.1.2 Cloud Absorption Model

We normally encounter different types of clouds. Their generation and formation mechanisms are different. The thunderstorm clouds affect the radio wave in a major way. This type of cloud sometimes extends well above the cirrus type and penetrates a few thousand feet into the stratosphere. These clouds must contain exceedingly high vertical velocities, and in that case the presence of hail above the tropopause is also possible. Sometimes it is also possible that under steady-state conditions the mass of cloud water per unit volume may be two to three times that of rain in the zone just below the melting level, especially in the case of light precipitation. This factor must be accounted for in radio wave propagation. The water content of clouds normally increases to a maximum in the vicinity of the melting level, and then above a temperature level of 1 or 2°C, gradually decreasing to zero. However, such steady-state conditions do not exist with respect to water distributions in the cloud when updrafts equal or exceed the fall velocity of particles as rain. Such updrafts exist locally for periods of about 5 to 15 minutes in thunderstorms or other convective activities, and hence can lead to a high local concentration of water.

Now, we confine ourselves to nonprecipitating cloud liquid water drops of various sizes. For sufficiently smaller particles, the Rayleigh approximation can be used to calculate scattering and absorption coefficients. The total absorption coefficient or scattering can be obtained by integration over the size distribution of the particles. But the complex dielectric constant of the particle manifests as an important property for calculating the absorption, and it is well described by the dielectric relaxation spectra of Debye (1929). This relaxation frequency is strongly temperature dependent, which in turn depends on the viscosity of liquid water. But while calculating absorption for nonprecipitating cloud liquid water, we assume the Rayleigh absorption, for which the liquid absorption depends only on total liquid content and does not depend on drop size distribution and scattering is negligible. This assumption is valid when scattering parameter

$$\beta = \mathrm{mod}\left\{ n\left(\frac{2\pi r}{\lambda}\right)\right\} \ll 1$$

where, r is the particle radius, λ is the free space wavelength, and n is the complex refractive index.

2.4.1.3 Oxygen Absorption

Recall that oxygen is a diatomic molecule, and it interacts with the electromagnetic radiation through magnetic dipole transitions between fine structures. This transition gives rise to a single line at 118.75 GHz and a complex band between 50 and 70 GHz. In that band 33 of which have intensities greater than 0.8×10^{-17} cm^{2}Hz. According to Van Vleck (1947a), an oxygen molecule has a magnetic moment equal to two Bohr magnetrons. This permanent magnetic moment is the cause of paramagnetism observed in oxygen, and it also allows the molecule to couple to the magnetic field of electromagnetic waves. It has one unit of spin angular momentum. This unit of spin angular momentum perturbs the rotational states since it is coupled to the rotational motion of the molecules. The unit electron spin has three spatial rotational motions with respect to given rotational angular momentum vector, $\hat{K}$, so that each rotational level is split into three states $\hat{J} = \hat{K}+1, \hat{K}, \hat{K}-1$. Each $\hat{J}$ state of this so-called ρ type triplet arises from a different orientation of the spin with respect to the rotational motion of the molecule. The energy difference between successive values of $\hat{J}$ in any of these triplets is about 2 cm^{-1}, with a single exception, $\hat{J} = 0 \rightarrow 1$. Selection rules for magnetic dipole transitions allow induced transitions between these successive members of the triplet. Thus, for each value of rotational angular momentum quantum number there are two absorption frequencies in the 2 cm^{-1} region. This population of various rotational lines follows Boltzmann statistics, which show that the most populated state at room temperature is that one for which $\hat{K} = 13$. States with $\hat{K} = 25$ also have a significant population. The O^{16} molecule has zero nuclear spin angular momentum, so that symmetry considerations demand $\hat{K}$ should have only odd values. Thus, with two absorption lines for each $\hat{K}$, there are 25 lines that overlap at atmospheric pressure. This contributes significantly to the absorption in the 2 cm^{-1} region (Strandberg et al., 1948).

At low pressure, there is a smaller oxygen concentration, leading to fewer collisions between the molecules. In fact, the width of collision-induced spectrum depends upon the number of collisions. With the occurrence of fewer collisions, individual resonance frequencies are distinguishable and result in a smaller attenuation due to atmospheric oxygen (Valdez, 2001). In fact, the microwave transitions are broadened over a range of frequencies by the effects of molecular collisions, which perturb the energy levels between the allowed transition levels. It should be noted that the collisions between like molecules, e.g, oxygen-oxygen collisions, produce greater broadening (self-broadening) in general than those between dissimilar molecules, such as oxygen-nitrogen collisions. This transition gives rise to a complex band between 50 and 70 GHz at sea level. Collisional broadening merges individual lines into an interacting band that has been an area of research for several decades. As the pressure increases, the width of individual resonance lines increases and contributes to neighboring line frequencies, leading to a higher total attenuation centered at about 60 GHz (Liebe, 1985a). The behavior of atmospheric oxygen is thus frequency and pressure and partially temperature dependent. It also shows inclusion of different model parameters that results in an increase of 20% accuracy from the entire previous models. But it is to be mentioned that Gasiewski and Staelin (1990) has shown a discrepancy of about 15% between the measured absorption by using a microwave temperature sounder (MTS) and the calculated absorption around the tropopause. As the height increases, atmospheric pressure reduces, and so too the rate and number of molecular collisions, resulting in less perturbation of energy levels. Consequently, the transitions are broadened over a smaller range of frequencies. Thus at low pressures, the individual transitions become narrower and start to become resolved. Furthermore, the transitions are not completely independent and do interact with each other. This causes the overall attenuation at a given frequency to be somewhat less than that expected from a simple summation of the effects of individual transitions.

A plot (Figure 2.1) generated for atmospheric oxygen attenuation using an empirical model developed by Liebe (1985b) shows the oxygen spectrum over Kolkata at an altitude of 9.67 km equivalent to a pressure of 300 hPa and air temperature of 245.88 K. It is clear from the Figure 2.1 that some of the individual transitions are beginning to resolve. It should be mentioned that this model is used because of the representative nature of the computationally efficient line-by-line model of atmospheric attenuation. Here atmosphere is treated locally as an isotropic gas at thermal equilibrium and the anisotropy introduced by the Zeeman effect is ignored. For typical (mid-latitude) terrestrial magnetic field strength, the maximum Zeeman splitting of the O_2 spectral line is of the order of 1 MHz, while collisional broadening is proportional to air pressure with a constant of proportionality of the order of 1 MHz/mb (Rosenkranz, 1993). At about 20.84 km altitude, with a pressure of 50 hPa and temperature around 220 K, most of the transitions are clearly resolved, as is evident from Figure 2.1. This suggests that the pressure-broadened absorption line decreases in its width as the altitude increases. It is also seen that pressure broadening is pressure and temperature dependent between 50 and 70 GHz.

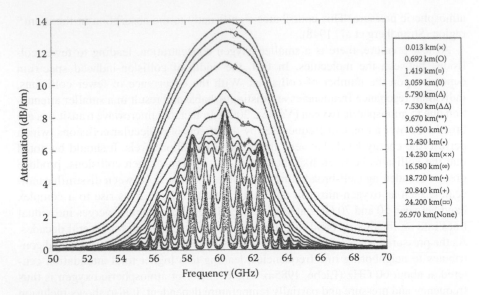

FIGURE 2.1 Microwave spectrums in the oxygen band in Kolkata, drawn by using MPM model.

2.4.1.4 Water Vapor Absorption

Theoretical estimates of the millimeter-wave attenuation due to atmospheric gases indicate that under clear weather conditions (no rain or fog), the attenuation rate in dB/km, also called the specific attenuation, exhibits minima around 35, 94, 140, and 220 GHz, which are called the millimeter-wave window, where the available bandwidths are of the order of 16, 23, 26, and 70 GHz, respectively (Button and Wiltse, 1981; Sen et al., 1986). In between these windows there exist the maxima of attenuation due to water vapor and oxygen molecules (Rogers, 1953, 1954; Straiton and Tolbert, 1960; Van Vleck, 1947a, 1947b). The maxima due to water vapor occur around 22.235 GHz and 183.311, while for oxygen the maxima occur at 61.151 and 118.75 GHz. A sample plot of the microwave spectrum by using the well-known millimeter-wave propagation model (MPM) by Liebe (1985) has been presented (Figure 2.2). Two representative places, such as Kolkata (22°N) and Srinagar (34°N), have been chosen. Kolkata is situated near the Bay of Bengal, and Srinagar is a hill city.

The second-order water vapor cluster, i.e., dimer, has resonance lines at 212.1 and 636.6 GHz (Viktorova and Zhevakin, 1976). Zenith attenuation from 100 to 1000 GHz is discussed by Emery and Zavody (1979).

The 22.235 GHz line is suitable mainly for ground-based study as well as monitoring, which at this frequency can provide valuable information regarding the total water vapor content, diurnal variation of water vapor content, and even the height distribution of water vapor with some simplifying assumptions. Such

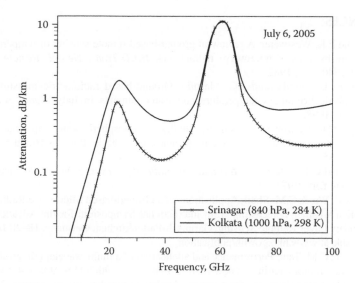

FIGURE 2.2 Microwave spectrums in the frequency band 10–100 GHz over Srinagar and Kolkata, India.

studies are useful not only for earth-space paths and horizontal links, but also for many moisture-related processes. These types of studies are important for cloud seeding experiments and for many meteorological research applications as well (Bhattacharya, 1985).

However, it is commonly believed that the spectral behavior of atmospheric constituents in the microwave and millimeter-wave band offers a good opportunity for the measurement of water vapor and liquid water also. The radiative properties of water vapor and liquid water are significantly different (refer to Figure 1.12), as shown by Karmakar et al. (2002).

Besides these, if we refer back to Figure 2.1, we see that the peak at 60 GHz remains invariant in magnitude (>10 dB/km), but on the other hand, the weak resonance peak occurring at 22.234 GHz shows a larger value over Kolkata than that over Srinagar. Hence, it is presumably considered that Kolkata always bears a little higher value of water vapor in comparison to that over Srinagar. Now, regarding the minima, we see that there occurs an appreciable shift of the conventional window frequencies from 30 and 94 GHz. It should be mentioned here that at a given frequency, the atmospheric absorption coefficient is a function of three basic parameters: temperature, pressure, and water vapor density. The water vapor absorption is essentially dependent on water vapor density, and likewise, the oxygen absorption is dependent on pressure and temperature. Since water vapor density and pressure decrease exponentially with increasing altitude, the major contribution in finding the zenith opacity (dB) is provided by the layers closest to the surface (Ulaby et al., 1981). Hence, it is expected that the zenith opacity will vary linearly with the surface water vapor density, which is supported experimentally by Waters (1976).

REFERENCES

Askne, J., and E.R. Westwater. A review of ground-based remote sensing of temperature and moisture by passive microwave radiometers. *IEEE Trans. Geosci. Remote Sensing*, GE-24, 340–352, 1982.

Barbaliscia, F., E. Fionda, and P.G. Masullo. Ground-based radiometric measurement of atmospheric brightness temperature and water contents in Italy. *Radio Sci.*, 33(3), 697–706, 1998.

Bhattacharya, C.K. Microwave radiometric studies of atmospheric water vapour and attenuation measurements at microwave frequencies. PhD thesis, Benaras Hindu University, India, 1985.

Button, K.J., and C.J. Wiltse, eds. *Infrared and millimeter waves*, 5. Vol. 4. Academic Press Inc., New York, 1981.

Chan P.W, K.C. Wu & C.M. Shun, Applications of a Ground-based Microwave Radiometer in Aviation Weather Forecasting, 13th International Symposium for the Advancement of Boundary Layer Remote Sensing, Garmisch-Partenkirchen, Germany, 18–20 July 2006. Available at www.hko.gov.hk/publication

Chan, P.W., and C.M. Tam. Performance and application of a multi-wavelength, ground-based microwave radiometer in rain now-casting. Presented at 9th IOAS-AOLS of AMS, 2005. Available at https://ams.confex.com/ams/Annual2005/techprogram/program_262.htm

Chandrasekhar, S., 1960, Radiative Transfer, Dover Publication, Inc, New York.

Clough, S.A., M.W. Shephard, E.J. Mlawer, M.J. Deamere, K. Cady-Pereira, S. Boukabara, and P.D. Brown. Atmospheric radiative transfer modelling: A summary of the AER codes. *J. Quant. Spectrosc. Radiative Transfer*, 9, 233–244, 2005.

Czekala, H., M.S. Crewell, C. Simmer, A. Thiele, A. Hornbostel, and A. Schroth. Interpretation of polarization features in ground-based microwave observations as caused by horizontally aligned oblate spheroids. *J. Appl. Meteorol.*, 40, 1918–1932, 2001.

Debye, P. *Polar molecules.* Dover, New York, 1929.

Decker, M.T., E.R. Westwater, and F.O. Guiraud. Experimental evaluation of ground-based microwave radiometric sensing of atmospheric temperature and water vapor profiles. *J. Appl. Meteorol.*, 17, 1788–1795, 1978.

Doran, J.C., S. Zhong, J.C. LIljegren, and C. Jakob. A comparison of cloud properties at a coastal and inland site at the North Slope of Alaska. *J. Geophys. Res.*, 107(D11), 4120, 2002. doi: 10.1029/2001JD000819.

Emery, R.J., and A.M. Zavody. Atmospheric propagation in the frequency range 100–1000 GHz. *Radio Electronic Eng.*, 49(7/8), 370–380, 1979.

Gasiewski A. J and D. H. Staelin, Numerical modeling of passive microwave O_2 observations over precipitation, Radio Science, 25, 3, 217–235, May June 1990.

Gasiewski, A.J. Microwave radiative transfer in hydrometeors. In *Atmospheric remote sensing by microwave radiometry*, ed. M.A. Janssen. Wiley, New York, 1993.

Goody, R.M., and Y.L. Yung. *Atmospheric radiation: Theoretical basis.* 2nd ed. Oxford University Press, Oxford, 1995.

Guiraud, F.O., J. Howard, and D.C. Hogg. A dual-channel microwave radiometer for measurement of precipitable water vapor and liquid. *IEEE Trans. Geosci. Electronics*, GE-17, 129–136, 1979.

Janssen, M. *Atmospheric remote sensing by microwave radiometry.* John Wiley, New York, 1993.

Karmakar, P.K., M. Maiti, S. Chattopadhyay, and M. Rahaman. Effect of water vapour and liquid water on microwave absorption spectra and its application. *Radio Sci. Bull.*, 303, 32–36, 2002.

Leuskii, V., V. Irsov, E. Westwater, L. Fedor, and B. Patten. Airborne measurements of the sea air temperature difference by a scanning 5 mm wavelength radiometer. In *Proceedings of IDRASS 2000*, Honolulu, HI, 2000, pp. 24–28.

Liebe, H.J. An updated model of millimetre wave propagation in moist air. *Radio Sci.*, 20(5), 1069–1089, 1985a.

Liebe, H.J. An atmospheric millimeter wave propagation model. *Int. J. Infrared Millimeter Wave*, 10(6), 631–650, 1985b.

Liebe, H.J., G.A. Hufford, and T. Manabe. A model for the complex permittivity of water at frequencies below 1 THz. *Int. J. Infrared Millimeter Waves*, 12(7), 659–675, 1991.

Liebe, H.J., and D.H. Layton. *Millimetre wave properties of the atmosphere: Laboratory studies and propagation modeling*. Report 87–24. National Telecommunications and Information Administration, Springfield, VA, 1987.

Liljegren, J.C., S.A. Boukabara, K. Cady-Pereiria, and S.A. Clough. The effect of the half width of the 22GHz water vapor line on retrieval of temperature and water vapour profiles with a twelve channel microwave radiometer. *IEEE Trans. Geosci. Remote Sensing*, 43(5), 1102–1108, 2005.

Liu, G.R., C.C. Liu, and T.H. Kuo. Rainfall intensity estimation by ground-based dual-frequency microwave radiometers. *J. Appl. Meteorol.*, 40, 1035–1041, 2001.

Marzano, F.S., C. Domenico, P. Ciotti, and R. Ware. Modelling and measurement of rainfall by ground-based multispectral microwave radiometry. *IEEE Trans. Geosci. Remote Sensing*, 43(5), 2005.

Marzano, F.S., E. Fionda, P. Ciotti, and A. Martellucci. Rainfall retrieval from ground-based multichannel microwave radiometers. In *Microwave radiometry and remote sensing of the environment*, ed. P. Pampaloni, 397–405. VSP, Utrecht, The Netherlands, 1999.

Marzano, F.S., E. Fionda, P. Ciotti, and A. Martellucci. Ground-based multi-frequency microwave radiometry for rainfall remote sensing. *IEEE Trans. Geosci. Remote Sensing*, 40, 742–759, 2002.

Rogers, T.F. Calculated centimeter-millimeter water vapour absorption at ground level. Presented at Proceedings of the Conference on Radio Meteorology, University of Texas, 1953.

Rogers, T.F. Absolute intensity of water vapour absorption at microwave frequencies. *Phys. Rev.*, 93, 248–249, 1954.

Rosenkranz, P.W. Absorption of microwaves by atmospheric gases. In *Atmospheric remote sensing by microwave radiometry*, ed. M.A. Janssen. Wiley, New York, 1993.

Rosenkranz, P.W. Water vapour microwave continuum absorption: A comparison of measurements and models. *Radio Sci.*, 33(4), 919–928, 1998.

Rosenkranz, P.W. Correction to water vapour microwave continuum absorption: A comparison of measurements and models. *Radio Sci.*, 34(4), 1025, 1999.

Sen, A.K., A.K. Devgupta, P.K. Karmakar, A. Mitra, and S.N. Ghosh. *Millimeter wave propagation in clear weather*. Report RPE-2. Electronics Commission, India, 1986.

Sheppard, B.E. Effect of rain on ground-based microwave radiometric measurements in the 20–90 GHz. *J. Atmos. Oceanic Technol.*, 13, 1139–1151, 1996.

Shih, S.-P., and Y.-H. Chu. Studies of 19.5 GHz sky radiometric temperature: Measurements and applications, *Radio Sci.*, 37(3), 1030, 2002. doi: 10.1029/2000rs002596.

Snider, J.B. Ground-based sensing of temperature profiles from angular and multi-spectral microwave emission measurements. *J. Appl. Meteorol.*, 11, 958–967, 1972.

Snider, J.B., F.O. Guiraud, and D.C. Hogg. Comparison of cloud liquid content measurement by two independent ground-based systems. *J. Appl. Meteorol.*, 19, 577–579, 1980.

Straiton, A.W., and C.W. Tolbert. Anomalies in the absorption of radio waves by atmospheric gases. *Proc. IRE*, 48, 898–903, 1960.

Strandberg, M.W.P., C.Y. Meng, and J.G. Ingersoll. *The microwave absorption spectrum of oxygen*. Technical Report 87. Research Laboratory of Electronics, MIT, Cambridge, MA, 1948.

Tretyakov, M.Yu., M.A. Koshelve, V.V. Dorovskikh, D.S. Makarov, and P.W. Rosenkranz. 60 GHz oxygen band: Precise broadening and central frequencies of fine structure, absolute absorption profile at atmospheric pressure, and revision of mixing coefficients. *J. Mol. Spectrosc.*, 231, 1–14, 2005.

Ulaby, F.T., R.K. Moore, and A.K. Fung. *Microwave remote sensing—Active and passive*, 2. Artech House, Norwood, MA, 1986.

Ulaby, F.T., R.K. Moore, A.K. Fung, et al. *Microwave remote sensing: Active and passive: Microwave remote sensing fundamentals and radiometry*. Vol. 1, no. 2. Addison-Wesley, Reading, MA, 1981.

Valdez, A.C. MS thesis, *Analysis of Atmospheric Effects Due to Atmospheric Oxygen on a Wideband Digital Signal in the 60 GHz Band*. Virginia Polytechnic Institute and State University, Blacksburg, 2001.

Van Vleck, J.H. The absorption of microwaves by oxygen. *Phys. Rev.*, 71, 1947a.

Van Vleck, J.H. The absorption of microwaves by uncondensed water vapour. *Phys. Rev.*, 71(7), 425–433, 1947b.

Van Vleck, J.H., and V.F. Weisskopf. On the shape of collision broadened lines. *Rev. Modern Phys.*, 17, 227–236, 1947.

Viktorova, A.A., and S.A. Zhevakin. Rotational spectrum of water vapour dimmers. *Radiophys. Quantum Electronics*, 18, 1976.

Ware, R., R. Carpenter, J. Guldner, J. Liljegren, T. Nehrkorn, F. Solheim, and F. Vandenberghe. A multichannel radiometric profiler of temperature, humidity, and cloud liquid. *Radio Sci.*, 38(4), 8079, 2003. doi: 10.1029/2002RS002856.

Waters, J.W. Absorption and emission of microwave radiation by atmospheric gases. In *Methods of experimental physics*, ed. M.L. Meeks, 2–3. Vol. 12, part B. Academic Press, New York, 1976.

Westwater, E.R. Ground-based passive probing using the microwave spectrum of oxygen. *J. Res. Natl. Bur. Stand.*, D69, 1201–1211, 1965.

Westwater, E.R., Crewell, C. Matzler, and D. Cimini. Principles of surface-based microwave and millimeter wave radiometric remote sensing of the troposphere. *Quaderni Della Societa Italiana Di Electromagnetsimo*, 1(3), 2005.

Westwater, E.R., and J.B. Snider, Ground based radiometric observation of atmospheric emission and attenuation at 20.6,31.65, and 90 GHz: A comparison of measurements and theory, IEEE Trans. Antenna and Propagation, 38, 10,1569–1580, 1990.

Westwater, E.R., and F.O. Guiraud. Ground based microwave radiometric retrieval of precipitable water vapour in the presence of clouds with high liquid content, *Radio Sci.*, 13(5), 947–957, 1980.

Westwater, E.R., J.B. Snider, and A.V. Carlson. Experimental determination of temperature profiles by ground-based microwave radiometry. *J. Appl. Meteorol.*, 14, 524–539, 1975.

Yang Meixue, Tandong Yao, Huijun Wang, Lide Tian, Xiaohua Gou, 2006, Estimating the criterion for determining water vapour sources of summer precipitation on the northern Tibetan Plateau, Wiley, Hydrological processes, 20, 3, 505–513.

3 Ground-Based Zenith-Looking Radio Visibility at Microwave Frequencies over a Tropical Location

3.1 INTRODUCTION

The ever-increasing demand of radar, communication, and navigational aids is apparently leading to an overcrowding of the electromagnetic spectrum. To meet this demand, the need for encroaching upon the higher frequencies, extending perhaps into the millimeter-wavelength region of the electromagnetic spectrum, is widely recognized. At such higher frequencies, however, we have to consider the important role where the electromagnetic interaction starts with the neutral atmosphere under various meteorological conditions. This particular phenomenon at frequencies above 10 GHz produces attenuation of the signal, polarization, time delay, etc.

It has already been mentioned that theoretical estimates of the millimeter-wave attenuation due to atmospheric gases produce the so-called minima around 35, 94, 140, and 220 GHz, which are called millimeter-wave windows, where the available bandwidths are of the order of 16, 23, 26, and 70 GHz, respectively (Button and Wiltse, 1981; Sen et al., 1986). But these frequencies occur in the U.S. standard atmosphere. The exceptions have already been discussed in Chapter 1. In between these windows there exist the maxima of attenuation due to water vapor and oxygen molecules (Rogers, 1953, 1954; Straiton and Tolbert, 1960; Van Vleck, 1947a, 1947b). The maxima occur around 22, 183, and 325 GHz due to water vapor and around 60,118 GHz due to oxygen. It should be mentioned that the maxima remain more or less invariant. At 22.235 GHz over Kolkata, the specific attenuation at the surface is 0.6 dB/km. This is predominantly due to large water vapor abundances at Kolkata (22°N). An experimental study by deploying a 22.235 GHz zenith-looking radiometer at Kolkata (Karmakar et al., 1999) supports this to a good proximity. In this case, initially it was assumed that the radiometer possessed radio visibility extended up to infinity. But this may not be the case in a place where water vapor seems to be a key factor in contributing toward absorption in the microwave band. A detailed discussion regarding water vapor absorption has been made by Karmakar (2011). But the oxygen spectra ranging from 50–70 GHz seem to be more complex and attenuation is sizable. Hence, it needs special attention, as this band is generally used as a temperature profiler. It also needs to explore the zenith-looking radio visibility, i.e., the height limit up to which there is no substantial increase of integrated attenuation in the zenith direction. It has been presumably assumed that while looking toward the zenith from the ground, a radiometer can look up to infinity. But that is not practically feasible. In order to study the height up to which ground-based radiometric attenuation takes approximately the saturation value, we define a term *radio visibility* at different microwave frequencies. We have defined the radio visibility as the height at which the variation of total attenuation is less than or equal to 1% of that of the immediate preceding slab, considering the thickness of each slab is equal to 10 m, and hence is the 99% radio visibility in the water vapor band. But on the other hand, while defining this radio visibility or height limit in the oxygen band, we have considered the slab thickness equal to 10 m. The percentage change in attenuation is considered to be 0.1% despite the fact that the attenuation is high in the oxygen band in comparison to the water vapor band.

3.2   ABSORPTION IN THE WATER VAPOR BAND

The maxima due to water vapor occur around 22.235 GHz and 183.311 GHz, while for oxygen the maxima occur at 61.151 and 118.75 GHz. A sample plot of the microwave spectrum by using the well-known millimeter-wave propagation model (MPM) by Liebe (1985, 1989) has been presented (Figure 3.1) over Kolkata (22°N).

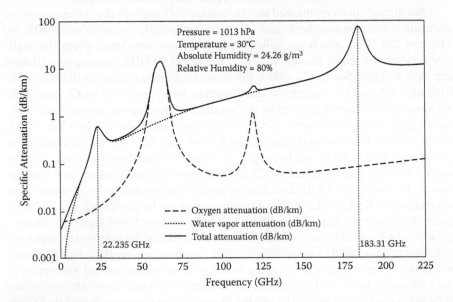

FIGURE 3.1   Electromagnetic wave attenuation profile due to water vapor and oxygen as a function of frequency in the millimeter-wave band over Kolkata.

The second-order water vapor cluster, i.e., dimer, has resonance lines at 212.1 and 636.6 GHz (Viktorova and Zhevakin, 1976). Zenith attenuation from 100–1000 GHz is discussed by Emery and Zavody (1979).

The 22.235 GHz line is suitable mainly for ground-based study. The continuous monitoring at this frequency is useful not only for earth-space paths and horizontal links, but also for many moisture-related processes (Deubey, 1980). These types of studies are important for cloud seeding experiments and for many meteorological research applications (Bhattacharya, 1985). However, it is commonly believed that the spectral behavior of atmospheric constituents in the microwave and millimeter-wave band offers a good opportunity for the measurement of water vapor and liquid water also. The radiative properties of water vapor and liquid water are significantly different (Figure 3.2), as shown by Karmakar et al. (2002).

Now, regarding the minima we see that there occurs an appreciable shift of the conventional window frequencies from 30–94 GHz. It was also observed that the minima are occurring at 75 GHz through 94 GHz over the globe during the months of January–February and 73–85 GHz during the months of July–August, depending on the latitudinal occupancy. It is observed that the large abundances of water vapor are mainly held responsible for shifting of minima toward the lower values of frequencies (Karmakar et al., 2011). It should be mentioned that

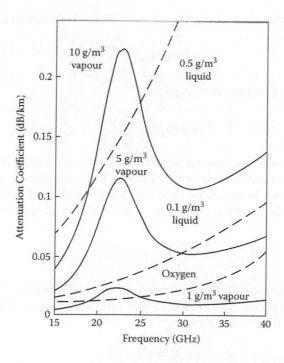

FIGURE 3.2 The atmospheric attenuation coefficient as a function of frequency for different densities of water vapor and liquid water.

at a given frequency, the atmospheric absorption coefficient is a function of three basic parameters: temperature, pressure, and water vapor density. The water vapor absorption is essentially dependent on water vapor density, and likewise, the oxygen absorption is dependent on pressure and temperature. Since water vapor density and pressure decrease exponentially with increasing altitude, the major contribution in finding the zenith opacity (dB) is due to the layers closest to the surface (Ulaby et al., 1981). Hence, it is expected that the zenith opacity will vary linearly with the surface water vapor density, which is supported experimentally by Waters (1976).

3.3 MEAN RADIATING TEMPERATURE

We consider a certain volume of the atmosphere that is considered to be an absorbing medium and attains a temperature T_m by absorbing the incident radiation from any external source. This in turn reradiates isotropically. The extent of such absorption or emission of energy depends on fractional transmissivity, σ, of the atmospheric medium. Thus, the radiated energy from the atmosphere is a noise that enhances the thermal noise temperature by an amount $(1 - \sigma)T_m$. If such an increase in thermal noise temperature is presented by T_a, then we write

$$T_a = (1 - \sigma)T_m \quad \text{Kelvin} \tag{3.1}$$

Again, by definition, the excess attenuation A (dB) is related to σ by

$$A = 10 \, log_{10}\left(\frac{1}{\sigma}\right) \tag{3.2}$$

which according to Allnut (1976) is given by

$$A = 10 \, log_{10}\frac{T_m - T_c}{T_m - T_a} \tag{3.3}$$

Here T_c is the cosmic background temperature and is assumed to be around 3.0 K in the microwave band. It might be noted that T_m is the function of frequency with a significant variation in microwave band (30–300 GHz). According to Altshuler and Lamers (1968), the empirical relation for $T_m(K)$ involving ground temperature T_s is given by

$$T_m = 1.12 \, T_s - 50 \tag{3.4}$$

Regression analyses between T_m and surface temperature T_s for different microwave frequencies in the range 22–140 GHz (Mitra et al., 2000; Karmakar, 2011) reveal the linear relation as

$$T_m = CT_s + D \tag{3.5}$$

The values of the constants are given in Table 3.1 for the sake of completeness.

TABLE 3.1
Best-Fit Linear Regression Coefficients for $T_m = CT_s + D$

Frequency (GHz)	Slope A (Kelvin/°C)	Intercept B (Kelvin)	Correlation Coefficient
22.235	0.823	267.383	0.986
31.4	0.857	267.354	0.985
53.75	0.819	269.045	0.982
67.8	0.864	266.800	0.987
76.0	0.911	266.691	0.983
94.0	0.928	268.246	0.988
118.75	0.966	266.975	0.982
120.1	1.004	265.57	0.979
125.0	0.998	267.288	0.985

3.4 WATER VAPOR CONTENT AND MICROWAVE ATTENUATION IN THE WATER VAPOR BAND

According to Buck (1981) and Karmakar et al. (2011), the theoretical relationships in finding the water vapor pressure, and hence water vapor density, are given by Equations 1.1 through 1.3.

Now by using the above equations, one can find the water vapor density at different heights, as available, and the integrated water vapor content in a cylinder of infinite height with a base equal to 1 m² (Figure 3.3) by integrating the functional dependence of density within height limit $h = 0$ to $h = \infty$.

While calculating the microwave specific attenuation (dB/km) at different heights, the input parameters were taken as pressure, ambient temperature, and dew point temperature, from which we can have the values of partial pressure of vapor and water vapor density, and hence the attenuation coefficient can be derived by using the well-known MPM model as prescribed by Liebe (1989). The attenuation coefficient at different heights up to 8 km for the month of August over Kolkata (22°N) is presented in Figure 3.4. It should be noted that the month of August has been chosen because this is the maximum vapor content-bearing month over Kolkata (Karmakar et al., 1999). It has also been assumed that the radiosonde observations would not be affected due to wind shear in the upper part of the troposphere. Now if we look at Figure 3.3, it is clear that the vapor content during the whole of the year does not vary appreciably beyond 7–8 km over Kolkata. For this reason, the specific attenuation has been calculated up to 8 km height, beyond which there is little possibility of getting traces of water vapor. This idea was also supported by Evans and Hagfors (1968).

Now to find out the total integrated attenuation (dB), a line-by-line summation may be adopted considering the atmospheric slab of 10 m thickness, wherein no considerable variation of vapor content has been taken place. Adopting this methodology, one can find out the integrated attenuation (dB) for 22, 31, 94, and 183 GHz over Kolkata. This is presented in Figure 3.5.

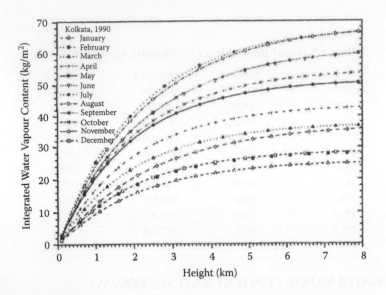

FIGURE 3.3 Variation of integrated water vapor content of the atmosphere as a function of height at Kolkata.

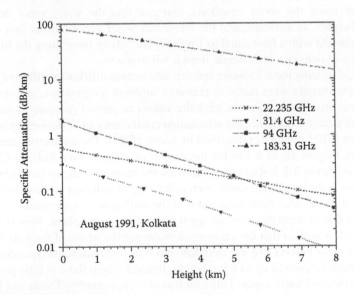

FIGURE 3.4 Variation of specific attenuation with height at a few selected frequencies in the millimeter-wave band.

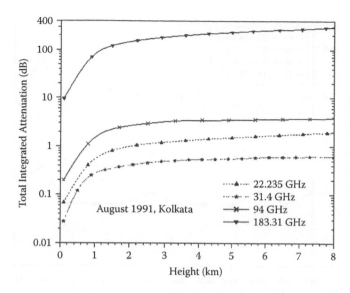

FIGURE 3.5 Variation of total integrated attenuation with height at a few selected frequencies in the millimeter-wave band.

3.5 DETERMINATION OF HEIGHT LIMIT OF RADIO VISIBILITY IN THE WATER VAPOR BAND

In finding the height limit, an effort has been made to calculate the percentage change of attenuation in dB between two immediate successive levels (each of 10 m thick) of the atmosphere from ground. This process is continued until the limit of 1% change of integrated attenuation is attained for a definite height limit for a given frequency. It has been found that over Kolkata, the height limit is 4.7 km for 22.235 GHz, and those for 31 GHz and 94 GHz are 4.3 and 3.8 km, respectively. On the other hand, this height limit at 183.31 GHz is found to be 4.9 km. Monthly variations of this 1% height limit at the selected frequencies are presented in Figure 3.6. This height limit has been restricted to 1% because no significant change in attenuation at different microwave frequencies in the water vapor band with respect to total attenuation is observed. So beyond this 1% height limit of attenuation, the radio receivers would not be able to have their appreciable radio vision. In other words, it may also be said that the aforementioned microwave frequencies over a tropical location like Kolkata would possesses 99% radio visibility at their corresponding height limits, as presented in Figure 3.6.

3.6 ABSORPTION IN THE OXYGEN BAND

Microwave and millimeter-wave radiation interact with diatomic oxygen in the atmosphere through magnetic dipole transitions between fine-structure spin-rotational levels of oxygen's electronic-vibrational ground state. These transitions give rise to a single line at 118.75 GHz and a complex band between 50–70 GHz, amongst

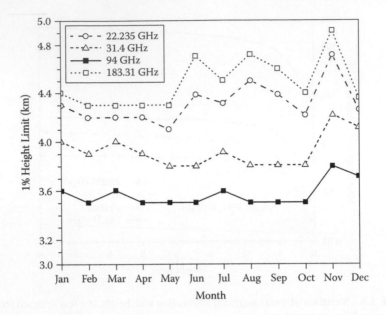

FIGURE 3.6 Monthly variation of 1% height limit at different microwave/millimeter-wave frequencies at Kolkata for the year 1991.

the lines present in the continuum, 33 of which have intensities greater than 0.8×10^{-17} cm²Hz. Around 60 GHz and at tropospheric pressure, the shape of these spectral lines is such that collisional broadening merges individual lines into an interacting band. In this regard, the MPM model of Liebe et al. (1992) may be used because of the representative nature of its computationally efficient line-by-line model of atmospheric absorption. Here, atmosphere is treated locally as an isotropic gas at thermal equilibrium and the anisotropy introduced by the Zeeman effect is ignored. For typical (mid-latitude) terrestrial magnetic field strength, maximum Zeeman splitting of the O_2 spectral line is of the order of 1 MHz, while collisional broadening is proportional to air pressure with a constant of proportionality on the order of 1 MHz/mb (Rosenkrantz, 1993).

The spectral line broadening and shifting that result from collisions among molecules in a gas are usually studied in the regime where observed effects are proportional to the frequency of isolated, binary collisions, and thus to gas pressure (Fano, 1963). As a result, the terms *pressure broadening* and *collisional broadening* are often used interchangeably, and broadening and shifting of lines are expected to be proportional to pressure. Numerous reviews of pressure broadening theory are available (Rosenkrantz, 1993; Breene, 1959; Ben-Reven, 1976; Levy et al., 1992) in this context. However, the attenuation coefficient (dB/km) in the oxygen band over Kolkata based on the MPM model is presented in Figure 2.1. There it is seen that in clear weather, the distinct transition starts to occur at about 300 mb pressure in the range 57–63 GHz band. This sharp transition in this band is useful for communication. Thus, the oxygen absorption coefficient is strongly dependent on partial pressure of air and has a uniform and time-invariant mixing ratio that favors the choice of oxygen band for temperature retrieving purposes (Ulaby et al., 1981). According to Westwater and Grody (1980), the vertical profile of the weighting function of 55.45 GHz decreases more rapidly with increasing height than the lower frequency, such as 52.85 and 53.85 GHz profiles. This implies that the surface layers within 1 km have greater influences on 55.45 GHz and the least influence on 52.85 and 53.85 GHz. Also at 55.45 GHz, it is dominated by the first 3–4 km of the lower atmosphere. The magnitude of the weighting function at 52.85 GHz remains significant up to 10 km height and even higher.

Now considering atmospheric pressure, ambient temperature, and dew point temperatures at different altitudes as the input parameters, the attenuation coefficients were calculated for Kolkata (urban). For this purpose, the well-known MPM model (Liebe, 1989) was used. The representative plots for 50, 53, 56, 60, and 65 GHz frequencies are given in Figure 3.7.

As discussed earlier, the atmosphere is initially considered to be a sum of 10 m thick slabs. A line-by-line summation is adopted to find out the total attenuation (dB) in the same frequency band. It is to be mentioned that the integration of the specific attenuations were computed over Kolkata for the year 2008. Now the integrated attenuation in dB over Kolkata is presented in Figure 3.8.

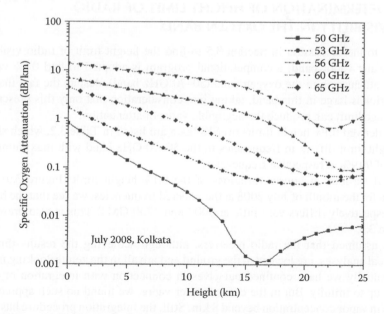

FIGURE 3.7 Height profile of specific oxygen attenuation at a few selected frequencies in the millimeter-wave band over Kolkata for the month of July 2008.

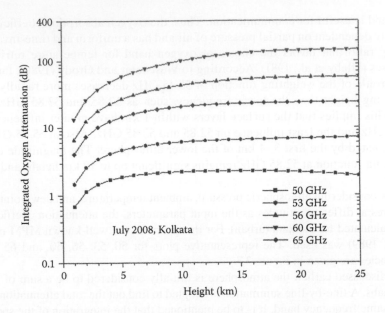

FIGURE 3.8 Variation of integrated oxygen attenuation with height at a few selected frequencies over Kolkata for the month of July 2008.

3.7   DETERMINATION OF HEIGHT LIMIT OF RADIO VISIBILITY IN THE OXYGEN BAND

Similar to the discussion in Section 3.5 to find the height limit of radio visibility in the water vapor band, a computational program is adopted to find 0.1% values of total attenuation in the oxygen band (50–70 GHz band), despite the fact that the attenuation is large in this band, taking into consideration that only this prescribed percentage limit can produce the negligible value of attenuation.

The derived 0.1% height limits over Kolkata are listed in Table 3.2, which shows the height limit differs in frequencies in the 50–70 GHz band with maximum visibility of 99.9% at respective frequencies.

But if we look at hourly variations of the 0.1% height limit (Figure 3.9) over Kolkata for the month of July 2008 at the selected frequencies, we see that the height limit respectively differs very little at 0000 and 1200 GMT. This is also presented in Table 3.3.

It is assumed that the radio receivers, although achieving the results through numerical analyses, are located at the ground and are all in the zenith-looking mode. So essentially we have confined ourselves in connection with integration or summation up to infinity. But in the case of water vapor, we found no such appreciable change in vapor concentration beyond 8 km. Still, the integration procedure has been made up to infinity to look into any height limit beyond the troposphere. This numerical technique may be considered to any extent of height where data are available, from balloon-borne or airborne experiments.

TABLE 3.2

Variation of 0.1% Height Limit over Kolkata at Different Microwave Frequencies for the Month of July 2008

Place	Kolkata				
Frequency (GHz)	50	53	56	60	65
0.1% height limit (km)	4.37	5.12	6.94	7.71	5.86

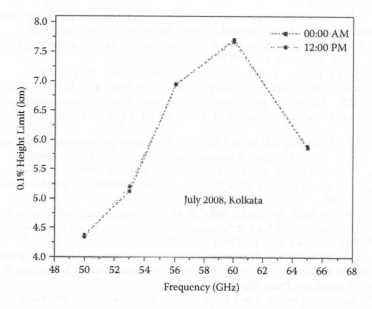

FIGURE 3.9 Daily variation of 0.1% height limit at different microwave frequencies over Kolkata for the month of July 2008.

TABLE 3.3

Hourly Variation of 0.1% Height Limit at a Few Selected Frequencies at Kolkata for the Month of July, 2008

Time	0530 GMT					1730 GMT				
Frequency (GHz)	50	53	56	60	65	50	53	56	60	65
0.1% height limit (km)	4.37	5.1	6.94	7.71	5.86	4.32	5.2	6.95	7.67	5.89

REFERENCES

Allnut, J.E. Slant path attenuation and space diversity results using 11.6 GHz radiometers. *Proc. IEEE*, 123, 1197–1200, 1976.

Altshuler, E., and U.H.W. Lamers. A troposcatter propagation experiment at 15.7 GHz over 500 km path. *Proc. IEEE*, 56, 1729–1731, 1968.

Bhattacharya, C.K. Microwave radiometric studies of atmospheric water vapour and attenuation measurements at microwave frequencies. PhD thesis, Benaras Hindu University, India, 1985.

Breene, R.G. *The shape of spectral lines*. Pergamon Press, New York, 1959.

Buck, A.L. New equations for computing vapour pressure and enhancement factor. *J. Appl. Meteorol.*, 20, 1527–1532, 1981.

Button, K.J., and J.C. Wiltse, eds. *Infrared and millimeter waves*, 5. Vol. 4. Academic Press, New York, 1981.

Deubey, V. Microwave radiometric studies of atmospheric water vapour and attenuation measurements at 22.235 GHz. PhD thesis, University of Delhi, India, 1980.

Emery, R.J., and A.M. Zavody. Atmospheric propagation in the frequency range 100–1000 GHz. *Radio Electronic Eng.*, 49(7/8), 370–380, 1979.

Evans, J.V., and T. Hagfors. *Radio astronomy*. McGraw Hill, New York, 1968.

Fano, U. Pressure broadening as a prototype of relaxation. *Phys. Rev.*, 131, 259–270, 1963.

Karmakar, P.K. *Microwave propagation and remote sensing: Atmospheric influences with models and applications*. CRC Press, Taylor & Francis Group, Boca Raton, FL, 2011.

Karmakar, P.K., S. Chattopadhyay, and A.K. Sen. Estimates of water vapour absorption over Calcutta at 22.235 GHz. *Int. J. Remote Sensing*, 20(13), 2637–2651, 1999.

Karmakar, P.K., M. Maiti, S. Chattopadhyay, and M. Rahaman. Effect of water vapour and liquid water on microwave absorption spectra and its application. *Radio Sci. Bull.*, 303, 32–36, 2002.

Karmakar, P.K., M. Maiti, S. Mondal, and C.F. Angelis. Determination of window frequency in the millimeter wave band in the range of 58° north through 45° south over the globe. *Adv. Space Res.*, 48, 146–151, 2011.

Levy, A., N. Lacome, and J.C. Chackerian. Collisional line mixing: Molecular spectroscopy. In *Atmospheric remote sensing by microwave radiometry*, ed. K.N. Rao and A. Weber, chap. 4. Academic Press, San Diego, 1992.

Liebe, H.J. An updated model of millimetre wave propagation in moist air. *Radio Sci.*, 20(5), 1069–1089, 1985.

Liebe, H.J. MPM—An atmospheric millimeter wave propagation model. *Int. J. Infrared Millimeter Waves*, 10, 631–650, 1989.

Liebe, H.J., P.W. Rosenkranz, and G.A. Hufford. Atmospheric 60-GHz oxygen spectrum: New laboratory measurements and line parameters. *J. Quantitative Spectrosc. Radiative Transfer*, 48, 1992.

Mitra, A., P.K. Karmakar, and A.K. Sen. A fresh consideration for evaluating mean atmospheric temperature. *Indian J. Phys.*, 74B, 379–382, 2000.

Reven, B. Spectral line shape in gases in the binary collision approximation. *Adv. Chem. Phys.*, 33, 1976.

Rogers, T.F. Calculated centimeter-millimeter water vapour absorption at ground level. Presented at Proceedings of the Conference on Radio Meteorology, University of Texas, November 1953.

Rogers, T.F. Absolute intensity of water vapour absorption at microwave frequencies. *Phys. Rev.*, 93, 248–249, 1954.

Rosenkrantz, P.W. Absorption of microwaves by atmospheric gases. In *Atmospheric remote sensing by microwave radiometry*, ed. M.A. Janssen. Wiley, New York, 1993.

Sen, A.K., A.K. Devgupta, P.K. Karmakar, A. Mitra, and S.N. Ghosh. *Millimeter wave propagation in clear weather*. Report RPE-2. Electronics Commission, India, 1986.

Straiton, A.W., and C.W. Tolbert. Anomalies in the absorption of radio waves by atmospheric gases. *Proc. IRE*, 48, 898–903, 1960.

Ulaby, F.T., R.K. Moore, A.K. Fung, et al. *Microwave remote sensing: Active and passive: Microwave remote sensing fundamentals and radiometry*. Vol. 1, no. 2. Addison-Wesley, Reading, MA, 1981.

Van Vleck, J.H. The absorption of microwave by oxygen. *Phys. Rev.*, 71(7), 413–424, 1947a.

Van Vleck, J.H. The absorption of microwaves by uncondensed water vapour. *Phys. Rev.*, 71(7), 425–433, 1947b.

Viktorova, A.A., and S.A. Zhevakin. Rotational spectrum of water vapour dimmers. *Radiophys. Quantum Electronics*, 18, 1976.

Waters, J.W. Absorption and emission of microwave radiation by atmospheric gases. In *Methods of experimental physics*, ed. M.L. Meeks, 2–3. Vol. 12, part B. Academic Press, New York, 1976.

Westwater, E.R., and N.C. Grody. Combined surface- and satellite-based microwave temperature profile retrieval. *J. Appl. Meteorol.*, 19, 1438–1444, 1980.

4 Radiometric Sensing of Temperature, Water Vapor, and Cloud Liquid Water

4.1 INTRODUCTION

A radio wave in the microwave/millimeter-wave band strongly interacts with ambient atmospheric particles like water vapor, oxygen, and hydrometeors. The effects of interaction are twofold. On the one hand, the atmospheric absorption, scattering, and refraction limit the performance of the microwave/millimeter-wave system, and on the other hand, this interaction allows the propagated wave to be used as the diagnostic tool for probing the lower atmospheric structure. Water vapor is perhaps the most important minor constituent that can affect the thermodynamic balance, photochemistry of the atmosphere, sun-weather relationship, and biosphere. The vertical and horizontal distribution of water vapor, as well as its temporal variation, is essential for probing into the mysteries of several effects. In fact, this triggered the need for the measurement of ambient water vapor for numerical weather prediction, short-term as well as severe storm forecasting. On the other hand, liquid water content measurement of clouds also provides very important input to the global circulation model (GCM).

It is well known that water vapor accounts for the largest percentage of the greenhouse effect, between 36 and 66% for water vapor alone, and between 66 and 85% when factoring in clouds. Water vapor concentrations fluctuate regionally, but human activity does not significantly affect water vapor concentrations except at a local scale, such as near an irrigated field. It has been discussed in Chapter 1 that the Clausius-Clapeyron relation establishes that air can hold more water when it warms. Hence, the basic principles indicate that warming associated with increased concentrations of the other greenhouse gases will also increase the concentration of water vapor, since water is a greenhouse gas and because warm air can hold more water.

Another important consideration is that with water vapor being the only greenhouse gas whose concentration is highly variable in space and time in the atmosphere, its real-time measurement is a highly emerging topic of research interest. The International Panel on Climate Change (IPCC) fourth assessment report (Cracknell and Varotsos, 2007) says that a further warming of about 0.1°C per decade would be expected even if the concentration of all greenhouse gases and aerosols had been kept constant. This report also says that in order to reduce the level of existing uncertainties, the modeling of the nature-society interaction is urgently required on a long-term basis, taking into account nonlinear changes in climate systems. The short-wave aerosols' radiative forcing at the surface in cloud-free conditions during the period 2000–2001 ranged from 10.8–20.1 W/m^2 in the winter and from 15.2–16.6 W/m^2 in the summer. The radiative heating rates near the surface due to aerosols were found to be in the range of 0.2–0.5 K/day during the winter period and 0.4 K/day during the summer period, simultaneous with enhanced heating in the lower troposphere (below 5 km). The long-wave radiative forcing (clear sky) at the top of the atmosphere induced by aerosols during nighttime was estimated to be only 0.02–0.04 and 0.04–0.05 W/m^2 for the winter and summer months, respectively (Tzanis and Varotsos, 2008). However, it is interesting to get a vivid idea of water vapor concentration over a particular place of choice. In this context, it is to be mentioned that the ground-based remote sensing in the microwave band has been known as a feasible means for the measurement of water vapor (Chong and Lu, 1994). In fact, the electromagnetic radiation at microwave frequencies interacts with the suspended atmospheric molecules, in particular with oxygen and water vapor. This interaction may be manifested in two ways in terms of complex refractivity. The imaginary part is generally expressed as attenuation and real part deals with propagation delay, which will be discussed in later in Chapter 7.

However, it is well known that the thermal emission around 20–30 GHz depends on both water vapor and liquid water. Thus, unless one restricts the problem to the clear sky condition only, one has to face the mathematical complexity, and in that case, the water vapor measurement will be influenced by an unknown amount of liquid (Sen et al., 1991). So, care has to be taken to separate the brightness temperature data for clear sky conditions only by recognizing the presence of overhead clouds with the help of an in-built infrared sensor in the radiometer itself, which computes the cloud base temperature (Karmakar et al., 2011) measurement. This measurement technique and the necessary algorithm for the single frequency have been developed by Karmakar (2011). In this context, the ground-based microwave radiometric sensing appears to be one of the suitable solutions for continuous monitoring of ambient atmospheric water vapor. Radiometric data have been extensively used by several investigators (Westwater, 1972, 1978; Grody, 1976; Westwater and Guiraud, 1980; Pandey et al., 1984; Janssen, 1985; Cimini et al., 2007; Karmakar and Chattopadhyay, 2004) to determine water vapor budget.

The first absorption maxima, although weak, in the microwave band occur at 22.234 GHz. So, one can have the choice of exploiting 22.234 GHz on the basis of assumption that the signal-to-noise ratio is largest at this frequency, provided the vertical profiles of pressure and temperature are constant (Resch, 1983). But this does not happen in practice. Pressure and temperature are highly variable parameters of the atmosphere. In this connection, Westwater (1978) showed that the frequency, independent of pressure, lies both way around the resonance line, i.e., 22.234 GHz. This single-frequency measurement of water vapor was done by Karmakar et al. (1999) at Kolkata (22°N) and at Instituto Nacional de Pesquisas Espaciais (INPE), Cachoeira Paulista (CP) (22°S), Brazil, (Karmakar et al., 2011a). But incidentally, 22.235 GHz is affected by pressure broadening, although the resonance peak occurs there, and these measurements were not devoid of any influence by the presence of overhead cloud liquid (Sen et al., 1990). It has been shown (Simpson et al., 2002) that

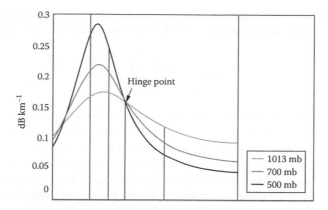

FIGURE 4.1 (See color insert.)   Representative plot of showing the occurrence of pressure independent frequency 23.834 GHz.

a zenith-pointing ground-based microwave radiometer measuring sky brightness temperature in the region of 22 GHz is three times more sensitive to the amount of water vapor than the amount of liquid water. However, in the region of 30 GHz, the sky brightness temperature is two times more sensitive to liquid water than that of water vapor, taking into consideration that the sensitivity to ice is negligible at both frequencies. Hence, it is suggested to use 23.834 GHz (hinge frequency) because of its independent nature of pressure broadening to get rid of unwanted signals in the vicinity of 23.834 GHz (Figure 4.1), along with 30 GHz for the measurement of water vapor budget (Karmakar et al., 2011a). But for profiling, it is always suggested to use the multifrequency measurement in the water vapor band (20–30 GHz).

However, radiosonde observations (RAOBs) are the fundamental method for atmospheric temperature, wind, and water vapor profiling, in spite of their inaccuracies, cost, sparse temporal sampling, and logistic difficulties. A better technology is the stable frequency-agile radiometric temperature and water vapor profilers, which can give continuous unattended profiles. They also have the capability to profile cloud liquid water, a capability absent in RAOBs and all other systems except for in situ aircraft devices. Applications for this passive radiometric profiling include weather forecasting (Karmakar et al., 1996) and now-casting, detection of aircraft icing and other aviation-related meteorological hazards, determination of density profiles for artillery and sound propagation, refractivity profiles for radio ducting prediction, correction to radio astronomy, satellite positioning and global positioning system (GPS) measurements, atmospheric radiation flux studies, estimation and prediction of telecommunication link degradation, and measurement of water vapor densities as they affect hygroscopic aerosols and smokes (Solheim et al., 1998).

4.2   GENERAL PHYSICAL PRINCIPLES

The basic idea of radiative transfer and thermal emission is given by Westwater (2004), and their application to microwave radiometric remote sensing is given by Goody and Yung (1995). In this regard, the radiometric principle was discussed Chapter 2.

According to Planck's law the spectral distribution of a black body emission, which may be expressed as the radiance $B_f(T)$ at temperature T and frequency f is given by

$$B_f(T) = \frac{2hf^3}{c^2} \frac{1}{e^{hf/KT} - 1} \tag{4.1}$$

Depending upon the emissivity of the target, the real body is sometimes called the grey body. If the fraction of incident energy from a certain direction absorbed by the grey body is $A(f)$, then the amount emitted is $A(f) B_f(T)$. For a perfectly reflecting or transmitting target, $A(f) = 0$ and incident energy may be reflected or transmitted without being absorbed in the body itself. For a vertically looking upward radiometer, the equation that relates the primary variable brightness temperature T_b to the radiative transfer equation is (Westwater et al., 2005)

$$B_f(T_b) = B_f(T)T_{\cos\mathrm{mic}} \exp\left\{-\int_0^\infty \alpha(h')\,dh'\right\} + \int_0^\infty B_f(T)T(h)\alpha(h)$$

$$\times \exp\left\{-\int_0^h \alpha(h')\,dh'\right\} dh \tag{4.2}$$

where h = path length in km, $T(h)$ = temperature (K) at the point h, and $\tau_v = \int_0^\infty \alpha(h)\,dh =$ total optical depth along the path h and $\alpha(h)$ = absorption coefficient (np/km) at the point h. The use of a black body source function in Equation 4.2 is justified by the assumption of local thermodynamic equilibrium, in which the population of emitting energy states is determined by the molecular collisions and is independent of the incident radiation field (Westwater, 2004). Equation 4.2 is used in several ways:

1. In forward model studies in which the relevant meteorological variables are measured by radiosonde soundings
2. In inverse problem and parameter retrieval applications in which meteorological information is obtained
3. In system modeling studies for determining the effects of instrument noise on retrieval applications

Figure 4.2 (Solheim et al., 1998) shows the absorption spectra induced by water vapor, oxygen, and cloud liquid water. Here absorption at two altitudes is shown to demonstrate pressure broadening.

4.3 THE FORWARD MODEL

We know that there exists a direct relation between the power radiated by an object and radiometric brightness temperature $T_B(f, \theta)$ of the scene under consideration. For an upward looking radiometer we may write, for nonscattering medium,

$$T_B(f,\theta) = T_{DN}(f,\theta) + T_{\cosmic}(f)e^{-\tau_f \sec\theta} \tag{4.3}$$

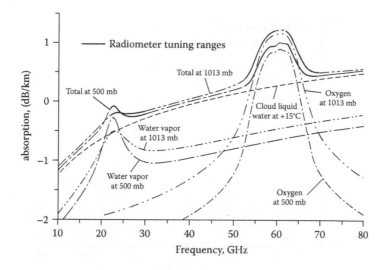

FIGURE 4.2 Atmospheric absorption by oxygen, water vapour and cloud liquid water is shown. Absorption at two altitudes is shown to represent pressure broadening. Radiometer tuning ranges are shown by broadened traces.

For all practical purposes, we may consider $T_{cosmic} = 2.7K$ (for $f \geq 10GHz$) sometimes it also may be neglected so long the first-order calculation is considered. However, the wide range of values of $T_{galactic}$ may pose a serious problem for the use of microwave radiometry for earth observations at frequencies below 1 GHz. This factor, together with the problem of interference from man-made radio sources and poor spatial resolution attainable with reasonable-size antennas, is responsible for the very limited use of frequencies below 1 GHz for radiometric earth observations. Here T_{DN} is due to the atmospheric radiation at frequency f down-welling at an angle θ.

For a nonscattering atmosphere in local thermodynamic equilibrium, we have the integral solution of the radiative transfer equation

$$T_{DN}(f,\theta) = \sec\theta \int_0^\infty k_f(h)T(h)e^{-\tau_f\sec\theta}\, dh \qquad (4.4)$$

where $T(h)$ is the thermometric temperature at a height h, and τ_f is the zenith optical thickness of the atmosphere up to height h and is given by

$$\tau_f(0,h) = \int_0^h k_f(h')dh'(\text{np}) \qquad (4.5)$$

Here K_f is the absorption coefficient at frequency f at a height of choice. Here it was assumed that the scattering coefficient is much smaller than the absorption. But in case of clouds, and other types of precipitation, we are to assume the total absorption, which is mainly the sum total of dry attenuation, nonprecipitating cloud liquid water, and rain attenuation (Karmakar et al., 2001). Then total attenuation is to be written as

$$\alpha(total) = \alpha_{dry} + \alpha_{cloud} + \alpha_{rain}$$

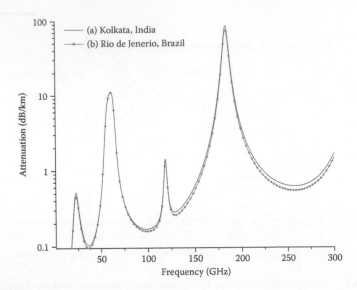

FIGURE 4.3 Microwave spectrum at Kolkata, India and Rio de Janerio, Brazil. Surface parameters for Kolkata are P = 1006 mb, T = 304.1 K and dew point temperature = 298.1K and those for Rio-de-Janerio are 996 mb, 304.1K and 299.7K.

In the absence of any precipitation and cloud, the gaseous absorption is equal to the total absorption. For a given set of atmospheric temperature, pressure, and water vapor density height profiles, $K_f(h)$ can be calculated for each height according to the Liebe model (1985) and expressed in dB-km^{-1}.

For the sake of clarity, the absorption coefficient in Kolkata (22°N), India, and Rio de Janeiro (23°S), Brazil (Karmakar et al., 2010), is presented (Figure 4.3). In the large attenuation regions, around a 60 GHz oxygen complex and the 118.75 GHz isolated line, the atmosphere behaves like a black body (Ulaby et al., 1981), with a radiometric temperature approximately equal to the weighted mean of the atmospheric temperature profile, where the weighting function accounts for the relative contribution of the various atmospheric layers to $T_{DN}(f, \theta)$. It is to be remembered that while calculating the absorption, the inputs are frequency f; the height profiles for atmospheric temperature $T(h)$; pressure P; and relative density $RH\%$. Now, to get $T_{DN}(f, \theta)$ using these parameters is called the forward problem (Ulaby et al., 1986).

4.4 THE INVERSE MODEL

Inverse problem deals with measured values of $T_{DN}(f, \theta)$ made at several frequencies and angles, as well as inputs to an inversion algorithm designed to produce estimates of one or more of the atmospheric variables as outputs. Since the forward problem is given by an integral equation, the inverse problem, such as finding the temperature and water vapor density profiles from the radiometric measurement of brightness temperature at the desired values of frequencies and angles, does not have a straight-forward solution. Hence, it is not unique and may be solved in principle through the application of the inversion technique.

4.4.1   INVERSION TECHNIQUES

We have already seen that the microwave spectrum of the atmosphere possesses several strong resonance lines due to oxygen and water vapor. Among these there exists a 60 GHz continuum that we generally call an oxygen continuum, and a 20–30 GHz band is exploited for water vapor and liquid water studies.

Now for formulating the inverse problem, let us assume that we are interested in remote sensing, i.e., retrieving the atmospheric temperature profile $T(h)$ from microwave ground-based zenith-looking radiometric measurement. In that case, we put $\theta = 0$ in Equation 4.4, and then it takes the form

$$T_{DN}(f) = \int_{0}^{\infty} W_{f,h} T(h)\, dh \tag{4.6}$$

where, $W_{f,h} = k_f(h)\, e^{-\tau(0,h)}$ and it is called the temperature weighting function, and $T(h)$ is the desired temperature profile. In practice, $T_{DN}(f)$ is usually measured at a discrete number of frequencies f_i and the objective of the inversion technique is to find a function $T(h)$ that, when substituted in Equation 4.6, will yield values of $T_{DN}(f_i)$ approximately equal to the measured values.

For U.S. standard atmosphere, the application of Equation 4.6 yields the plots (Figure 4.4) of normalized weighting function (normalized to unit maxima by dividing $W_{f,h}$ by its value at the surface) at the selected frequencies versus height (Westwater and Grody, 1980). Thus, $W_{f,0} = k_{f,0}$. If the integral of Equation 4.6 is approximated as the summation of m number of layers, each of height Δh, the observed radiometric temperature at frequencies f_i can then be written as

$$T_{DN}(f_i) = \sum_{j=1}^{m} W(f_i h_j) T(h_j) \quad i = 1,2\ldots\ldots,n \tag{4.7}$$

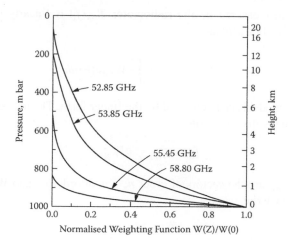

FIGURE 4.4   Height profile of normalized weighting function for US standard atmosphere in the 60 GHz continuum.

Now to avoid the mathematical complexity, let us write Equation 4.7 into a vector form as

$$\overrightarrow{T_{DN}} = \overline{W}\overrightarrow{T} \tag{4.8}$$

where $\overrightarrow{T_{DN}}$ and $\overrightarrow{T}$ are the vectors of dimensions n and m, respectively, and W is $(n \times m)$ matrix $(m > n)$. Apparently it seems that $W(f_i,h_j)$ depends only on frequency and height. But the weighting function depends on the absorption coefficient, which in turn depends on other atmospheric variables, like pressure $P(h)$, temperature $T(h)$, and water vapor density $\rho_v(h)$. Hence, $W(f_i,h_j)$ would change according to the change of all these parameters and pose a problem. To get rid of these, it is always suggested to use either multifrequency or multiangle observations. On the other hand, the counterbalance that renders the problem solvable is the *a priori* information about the character of the atmosphere for a given location and given time of year (Ulaby et al., 1986). Rodgers (1976, 1977) called this *a priori* information *virtual measurements*, which may include historical statistics about the profile, constraints imposed by atmospheric physics, and any other information that would narrow the range of values that $T(h)$ can have.

4.4.2 General Formulation

Now the main purpose is to get the solution of Equation 4.8. But before doing so, let us see how it would be more useful to consider these techniques in connection with forms that are somewhat more general than Equation 4.8. Basically, two forms of integral equations exist in the radiative transfer equation: linear and nonlinear. But the nonlinear form is basically applicable in the infrared region. So we will discuss the linear form only.

4.4.2.1 Linear Form

If the temperature measured by the radiometer $T_{DN}(f_i)$ is linearly related to the unknown guess function $g(h)$ as

$$T_{DN}(f_i) = T_{cosmic} + \int_0^\infty W_{f,h}g(h)\,dh \tag{4.9}$$

then the physical model is characterized by the linear form. The above equation can now be written in the discrete form as

$$\overrightarrow{T}_{DN} = T_{cosmic} + \overline{W}\,\vec{g} \tag{4.10}$$

Here T_{DN} and T_{cosmic} are the vectors of order n, g is of order m, and W is $n \times m$ the matrix. Let us now write

$$\overrightarrow{T}_m = \overline{W}\,\vec{g} \tag{4.11}$$

If we consider the contribution of T_{cosmic} to be constant or negligible, then Wg represents a vector and is considered the input measurement. Now our task is to generate the guess function depending on the historical statistics from which a mean profile or time average $<g(h)>$ can be generated. Now let the departure from the mean profile be

$$g'(h) \triangleq g(h) - <g(h)> \tag{4.12}$$

Now Equation 4.9 can be written in the form of the mean:

$$<T_{DN}(f_i)> = T_{cosmic} + \int_0^\infty W_{f,h} <g(h)> dh \tag{4.13}$$

From Equations 4.9 and 4.13, we write

$$T_{DN}(f_i) - <T_{DN}(f_i)> = \int_0^\infty W_{f,h} \{g(h) - <g(h)>\} dh = T'_{DN}(f_i),$$ and hence

$$T'_{DN}(f_i) = Wg' \tag{4.14}$$

The magnitudes of the errors strongly depend on the choice of frequencies.

4.4.3 VARIOUS RETRIEVAL METHODS

4.4.3.1 Optimal Estimation Method

This method needs sufficient *a priori* information using radiosonde observation. This in turn depends on historical background of the location and time of the year. Here the unknown vector g is of the order m using observations. One such form of the *a priori* information is the mean profile of $<g>$ and its covariance matrix S_g such that $S_g = <gg^T>$. For example, if someone is interested in getting the temperature profile, then the representative ensemble of radiosonde-measured temperature can provide $<g>$ and its covariance also.

4.4.3.2 Least Square Solution

As shown in Equation 4.11, we cannot measure the parameter's true form because of experimental errors, which include the measurement error and modeling errors, the latter being an approximation and assumptions underlying the model as given in Equation 4.11. Let us now denote the observations as $\hat{T}_m = T_m + \epsilon = Wg + \epsilon$, where ϵ is the error vector. The first criterion is to make at least $n \geq m$. The standard estimated solution of g of Equation 4.11 is provided by Franklin (1968) as

$$\hat{g} = (W^T W)^{-1} W^T \hat{T}_m \tag{4.15}$$

Here W^T is the transpose of W.

It is called the least square solution because it minimizes the sum of the squares of the differences between the measurements $\hat{T}_m$ and the values calculated from the model on the basis of the solution. But the least square solution is not well accepted for several reasons, as described by Ulaby et al. (1986).

4.4.3.3　Statistical Inversion Method

This method is perhaps the most widely accepted one in remote sensing. According to Westwater and Decker (1977), "Statistical inversion appears capable of extracting maximum information from the measurements." However, assuming that the observations $\hat{T}_m$ are linearly related to the unknown function g through

$$\hat{T}_m = Wg + \epsilon \tag{4.16}$$

and assuming further $<\epsilon> = 0$, we may write

$$\hat{T}'_m = Wg' + \epsilon \tag{4.17}$$

Here the primes denote the departure from mean values, i.e., $\hat{g}' = g' - <g>$ and

$$\hat{T}'_m = \hat{T}_m - <T_m> = \hat{T}_m - W<g> \tag{4.18}$$

Here the solution $\hat{g}'$ is related to the observations through $\hat{T}'_m$, a predictor matrix D through the relation

$$\hat{g}' = D\hat{T}'_m \tag{4.19}$$

This inversion method is closely related to the multiple regression method. Here D denotes the matrix whose elements are the regression coefficients. According to Franklin (1970) and Ishimaru (1978), the coefficients are given by

$$D = E\{g'(\hat{T}_m)^T\}E\{(\hat{T}'_m)(\hat{T}'_m)^T\} \tag{4.20}$$

Here E stands for expectation values. The value of D can be evaluated by minimizing the expected value of variance of the error in the estimate

$$E\{(g - \hat{g})^T(g - \hat{g})\} = \text{Minimum} \tag{4.21}$$

This is the reason for which this method is also called the minimum variance method.

In reality, if the measured radiometric $\hat{T}_m$ and corresponding values of g measured by radiosonde, can be made available, then the statistical quantities in Equation 4.20 also can be found out empirically by using standard regression programs. Thus, the element of D can be found without any knowledge of the weighting function.

4.4.3.4　Newtonian Iteration Method

The relationship between the measurement and the quantities to be retrieved may be expressed as

$$Y = FX \tag{4.22}$$

where Y is the m-dimensional measurement vector and X is the n-dimensional profile vector. It is important to note that in the problem, for a given measurement vector Y, an infinite number of profile vectors will satisfy Equation 4.22. Thus, a unique solution does not exist. Some additional information about the profile vector is required to constrain the solution. Here we need a large number of historical radiosonde profiles in the place of choice. The technique through which such information is incorporated is the Newtonian iteration method. According to Solheim et al. (1998), the $(k + 1)^{th}$ iteration solution can be expressed as

$$X_{k+1} = X_s + S_k K_k^T (K_k S_k K_k^T + S_e)^{-1} \{y - y_k - K_k (X_s - X_k)\} \qquad (4.23)$$

where $k = 0,1,2\ldots\ldots$, X_k is the kth solution, y is measurement vector with an error covariance matrix S_e and K_k is calculated as

$$K_k = \frac{\partial F}{\partial X} (\text{at } X = X_1) \qquad (4.24)$$

and contains weighting functions evaluated at the kth estimate X_k of X and

$$y_k = F(X_k) \qquad (4.25)$$

The statistical constraint is represented by X_s and S_k is the covariance matrix of the statistical ensemble.

4.4.3.5   Bayesian Maximum Probability Method

This model is based on Bayes' rule to estimate the most probable value P of the state vector x, for example, water vapor density from the observable vector y, like brightness temperature and surface meteorological data, as

$$P(a|y) = \frac{P(y|a)P(a)}{P(y)} \qquad (4.26)$$

The state vector that defines the temperature and vapor density profiles over a grid is represented as a Karhunen-Loeve expansion using eigenvectors derived from *a priori* information on the *covariance* of the state vector *a*. The state vector covariance matrix is calculated from representative radiosonde data. An advantage of the Karhunen-Loeve representation is that it can reduce the number of independent unknowns. If the eigenvalues of the *a priori* covariance matrix are ordered by decreasing value, it often happens that only a fraction is significant; the rest represent noise. The inversion problem then reduces to estimating a smaller set of variables, the computational burden is reduced, and the accuracy of the inversion can increase if the elements of the state vector covariance matrix are not well determined from the radiosonde archive (Klein, M. and Marsh 1996; Solheim et al., 1998). With the given set of variables the state vector is iterated and the corresponding theoretical observables are computed until the most probable, i.e., the maximization of Equation 4.26, is obtained. This minimizes the residuals between measured and computed observables.

4.5   RADIOMETRIC RESPONSE TO ATMOSPHERIC PROFILES: WEIGHTING FUNCTION

If we look at Equation 4.2, we find the theoretical radiometric response to atmospheric variables and absorption coefficient $\alpha(h)$, as a function of the path coordinate h. The absorption coefficient in turn is a function of temperature $T(h)$, pressure (h), water vapor density $\rho_V(h)$, and cloud liquid water density $\rho_L(h)$. Assuming that we are in the Rayleigh region, the response $B_f(T_b)$ to profiles is a nonlinear function of the above parameters. Thus, $B_f = B_f\{T, P, \rho_V, \rho_L\}$. The effect of water vapor and liquid water on microwave spectra is discussed by Karmakar et al. (2002). However, to get some idea of how small changes in the background meteorological profiles like $\delta T(h)$ bring a change in brightness, we write

$$\delta B(f) = \int_0^\infty \left\{ \left(W_T(h)\delta T(h) + W_P(h)\delta P(h) + W_\rho(h)\delta\rho_V(h) + W_L(h)\delta\rho_L(h) \right) \right\} dh \quad (4.27)$$

Here W's are the corresponding weighting functions of the atmospheric variables like temperature, etc. The detailed analytical expressions are provided in Chapter 2.

The height profiles of weighting functions at different places (see Table 4.1 for latitudinal occupancy) over the globe are presented in Figure 4.5(a–f).

Westwater and Snider (1990) presented the ground-based zenith-looking observation of atmospheric variables at 20.6, 31.65, and 90 GHz. At the locations of the experiments—San Nicolas Island, California, and Denver, Colorado—radiosonde observations of temperature and humidity were available. The data, after conversion to attenuation by the use of mean radiating temperature approximation, were processed to derive attenuation statistics. Both clear and cloudy attenuation characteristics were examined and compared with the results available from recent theories. For clear atmosphere, the water vapor model of Waters (1976) and that of Liebe (1989) were compared, and it was found that at 20.6 and 31.65 GHz, the model of Waters agreed better with the experiment, but at 90 GHz the model of Liebe was superior. A plot of weighting functions for zenith observation is shown in Figure 4.6(a–d), by Westwater and Snider (1990) for different atmospheric variables using Liebe's (1989) model for water vapor, Rosenkranz's (1988) model for oxygen, and Grant et al.'s (1957) model for cloud liquid. If we look at Figure 4.6(a), we see that for temperature

TABLE 4.1
Latitudinal Occupancy of Different Places

Places	Country	Latitude
Dumdum	India	22.65°N
Chongging	China	29.0°N
Aldan	Russia	58.0°N
Lima Calla	Peru	12.0°S
Porto Alegre	Brazil	29.0°S
Commodore	Argentina	45.0°S

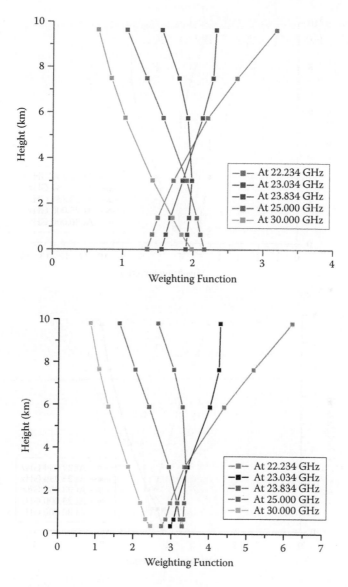

FIGURE 4.5 Height profile of weighting function a) Dumdum (22.65°N), India; b) Chongging (29°N), China.

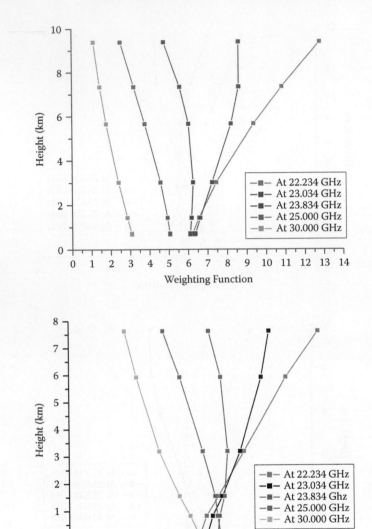

FIGURE 4.5 *(Continued)* Height profile of weighting function c) Aldan (58°N), Russia; d) Lima Calla (12°S), Peru.

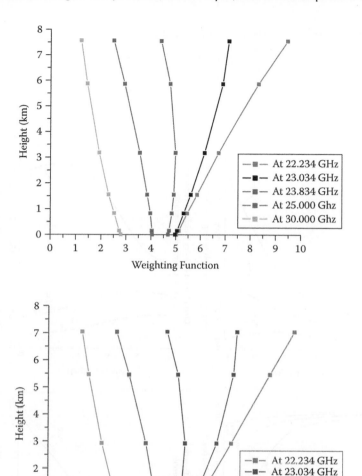

FIGURE 4.5 (Continued) Height profile of weighting function e) Porto Alegre (29°S), Brazil; f) Commodore (45°S), Argentina.

weighting function for 20.6 GHz, the average value for the temperature weighting function over the first kilometer is about −0.005/km. Thus, a change in temperature of 20K over the first kilometer would lead to a change in brightness temperature of about −0.1K. But relative to 20.6 and 31.65 GHz, the 90 GHz channel shows a greater sensitivity to liquid water density. But it is to be noted that below about 5 km, both vapor and cloud liquid weighting functions at 20.6 and 31 GHz are nearly constant with height. This implies that variation in brightness temperature is primarily affected by the variation in the column integrated amounts of vapor and liquid.

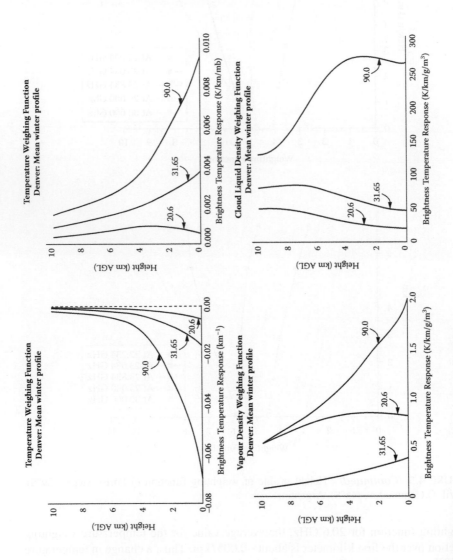

FIGURE 4.6 Weighting functions for ground based observation.

For calculating the weighting functions of different atmospheric variables radiosonde data may be subdivided into two classes: cloudy and noncloudy sets. Solheim et al. (1998) calculated weighting functions at 200 MHz intervals within the 20–29 (K-band) and 52–59 (V-band) GHz tuning wavebands and at 90,30 GHz and 14.5° elevation angles at Norman, Oklahoma, for the year 1992. This site was chosen because of its wide range of water vapor values and profile structures. Now, using the three different methods, namely, neural network, multiple regression, and Newtonian, the height profiles of temperature, water vapor, and cloud liquid are presented in Figure 4.7 (Soleihm, 1998).

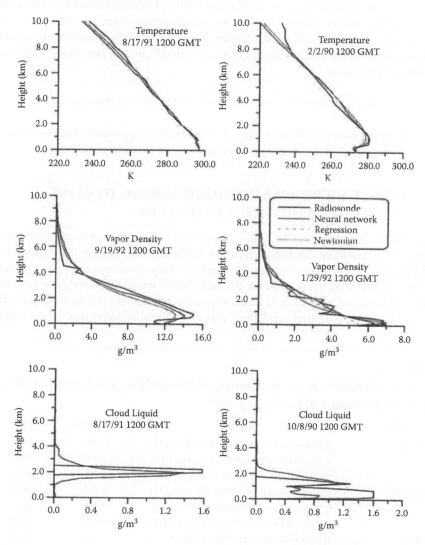

FIGURE 4.7   Retrieved profiles for cloudy radiosonde data sets from Norman, Oklahoma obtained by several methods as shown in the figure. The typical "good" results are on the left and "poorer" are on right.

4.6 PREDICTABILITY OF ATTENUATION BETWEEN VARIOUS FREQUENCIES

We already know from the microwave spectrum that for remote sensing purposes we should use multifrequency to retrieve the atmospheric parameters, especially in the case of a cloudy sky. But the higher the frequency, the more likely the possibility of not getting plentiful data. For example, an extensive amount of data in the water vapor band may be available all over the world. But, installation of another window frequency in a higher range, for example, 90 GHz, is costly and simultaneity may not be possible to maintain. In that case, it is of interest to examine between-channel predictability. Such considerations are important when attenuation is being estimated at various locations. Westwater and Snider (1990) examined regression relations between the various channels for clear, cloudy, and all conditions. The form of linear regression is

$$\tau, opacity, (dependent) = a_0 + a_1\tau_1 (independent) + a_2\tau_2 (independent).$$

The results are shown in Tables 4.2 to 4.4. But is to be remembered that this process of extrapolation is valid only for a particular location and season and time.

4.7 PASSIVE MICROWAVE PROFILING DURING DYNAMIC WEATHER CONDITIONS: A CASE STUDY

According to Knupp et al. (2012), the ground-based multichannel microwave radiometric observations reveal that at Boulder, Colorado, and Huntsville, Alabama, a short-period (1- to 5-minute) temperature and humidity soundings up to 10 km height are retrieved. But on the other hand, the radiometric retrievals provide substantially improved temporal resolution of thermodynamic profiles in contrast to radiosonde soundings. But the vertical resolution, which declines in proportion to the height

---

TABLE 4.2

Regression Relations between Absorption (dB) at 20.6, 31.65, and 90 GHz, San Nicolas Island, California, July 1987

Sky Condition	Relation	Correlation	Error
Clear sky	$\tau(20.6) = -0.71 - 0.055\tau(90) + 0.266\tau(31)$	0.986	0.011
	$\tau(31) = 0.086 + 0.141\tau(90) + 0.161\tau(20.6)$	0.995	0.003
	$\tau(90) = -0.332 - 0.107\tau(20.6) + 4.50\tau(31)$	0.991	0.015
Cloudy sky	$\tau(20.6) = -0.044 - 0.890\tau(90) + 5.72\tau(31)$	0.933	0.041
	$\tau(31) = 0.093 + 0.172\tau(90) + 0.089\tau(20.6)$	0.999	0.005
	$\tau(90) = -0.539 - 0.460\tau(20.6) + 5.73\tau(31)$	0.998	0.029
All data	$\tau(20.6) = -0.385 - 0.849\tau(90) + 5.43\tau(31)$	0.939	0.039
	$\tau(31) = 0.087 + 0.152\tau(90) + 0.092\tau(20.6)$	0.099	0.005
	$\tau(90) = -0.499 - 0.463\tau(20.6) + 5.64\tau(31)$	0.998	0.029

TABLE 4.3
Regression Relations between Absorption (dB) at 20.6, 31.65, and 90 GHz, Denver, Colorado, December 1987

Sky Condition	Relation	Correlation	Error
Clear sky	$\tau(20.6) = -0.272 - 0.023\tau(90) + 3.57\tau(31)$	0.889	0.024
	$\tau(31) = 0.076 + 0.136\tau(90) + 0.108\tau(20.6)$	0.972	0.004
	$\tau(90) = -0.396 - 0.252\tau(20.6) + 5.57\tau(31)$	0.958	0.026
Cloudy sky	$\tau(20.6) = 0.055 + 0.047\tau(90) + 0.443\tau(31)$	0.956	0.028
	$\tau(31) = 0.019 + 0.178\tau(90) + 0.455\tau(20.6)$	0.981	0.028
	$\tau(90) = -0.207 + 0.892\tau(20.6) + 3.27\tau(31)$	0.958	0.122
All data	$\tau(20.6) = 0.084 + 0.139\tau(90) + 0.107\tau(31)$	0.802	0.040
	$\tau(31) = 0.056 + 0.242\tau(90) + 0.014\tau(20.6)$	0.984	0.014
	$\tau(90) = -0.229 + 0.279\tau(20.6) + 3.78\tau(31)$	0.985	0.056

TABLE 4.4
Regression Relations between Absorption (dB) at 20.6, 31.65, and 90 GHz, Denver, Colorado, August 1987

Sky Condition	Relation	Correlation	Error
Clear sky	$\tau(20.6) = 0.039 - 0.451\tau(90) + 0.236\tau(31)$	0.968	0.027
	$\tau(31) = 0.056 + 0.177\tau(90) + 0.047\tau(20.6)$	0.961	0.012
	$\tau(90) = -0.148 + 0.072\tau(20.6) + 2.125\tau(31)$	0.978	0.043
Cloudy sky	$\tau(20.6) = 0.046 + 0.024\tau(90) + 0.425\tau(31)$	0.961	0.067
	$\tau(31) = -0.404 + 0.090\tau(90) + 0.995\tau(20.6)$	0.972	0.103
	$\tau(90) = -0.141 + 0.119\tau(20.6) + 3.46\tau(31)$	0.954	0.637
All data	$\tau(20.6) = 0.31 + 0.051\tau(90) + 0.487\tau(31)$	0.878	0.084
	$\tau(31) = 0.006 + 0.182\tau(90) + 0.151\tau(20.6)$	0.975	0.047
	$\tau(90) = -0.332 + 0.389\tau(20.6) + 4.347\tau(31)$	0.974	0.229

above ground level, is lower. Comparative analyses between radiometric retrievals and radiosonde soundings show that retrieval errors are similar in magnitude to radiosonde representativeness errors (<2 K in temperature, <1.5 g m^{-3} in water vapor density).

The high temporal resolution reveals detailed thermodynamic features of various rapidly changing hazardous weather phenomena. To illustrate the microwave radiometer profiler (MPR) capabilities and potential benefits to research and operational activities, some examples of radiometric retrievals from a variety of dynamic weather phenomena, including upslope super-cooled fog, snowfall, a complex cold front, a nocturnal bore, and a squall line accompanied by a low and other rapid variations in low-level water vapor and temperature, are presented.

The five different constrained examples are given as derived from the radiometric measurements.

4.7.1 Radiometric Measurements

4.7.1.1 Upslope with Super-Cooled Fog

Prolonged upslope weather conditions occurred in the Denver area from February 14 to 17, 2001. An initial cold frontal passage occurred early on February 14. A secondary cold front with a larger, colder, and more intense trailing anticyclone reinforced the upslope flow around Denver on February 16–17. Radiometric retrievals up to 2 km above ground level (AGL) (is about 1.5 km MSL) at Boulder on February 16, 2001, are shown in Figure 4.8. Advection of cold air behind the secondary cold frontal surge produced shallow moist upslope conditions along the Front Range, beginning near 1100 Coordinated Universal Time (UTC) on February 16. The passage of this front produced a modest 3°C decrease in surface temperature (left column, first plot), and a more significant 25% increase in surface relative humidity (left column, second plot) during the 1-hour period from 1100–1200 UTC. The radiometric retrievals in Figure 4.8 clearly show that the frontal passage was confined to levels below 1 km AGL, where the cold, moist air is most prominent.

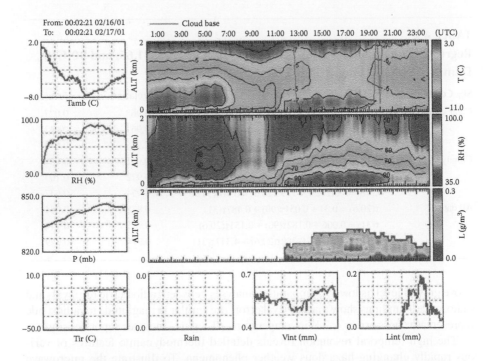

FIGURE 4.8 (See color insert.) Radiometric retrievals of a supercooled fog event associated with upslope flow at Boulder on 16 Feb 2001. Poor visibility and icy conditions during this upslope event led to major transportation disruptions including diversion of international flights from Denver for 18 hours. Time series of temperature (upper contour), humidity (middle contour), and cloud liquid density (lower contour) profiles are shown along with surface temperature (left row, top plot), humidity (left row, second plot), and pressure (left row, third plot); zenith infrared temperature (left column, bottom plot; rain flag (lower row, second plot); and integrated water vapor (lower row, third plot) and integrated liquid water retrievals (lower row, fourth plot).

Cloud liquid water density reached a maximum of 0.32 g m^{-3} at 300 km AGL at 1716 UTC, when the depth of the cold, moist air was near a maximum. Cloud base height, which is automatically retrieved from infrared measurements of cloud base temperature (left column, bottom plot) and the retrieved temperature profile, is indicated in the upper contour plot. The temperature profile indicates that the fog was super-cooled, with a minimum temperature near the top (~500 m AGL) of the stratus cloud deck of about −8°C. The rapid onset of fog behind the cold front is indicated by sharp increases in TIR (left column, bottom plot) the fourth month of the Iranian calendar and integrated liquid (bottom row, fourth plot), which increased to a maximum of 0.19 mm at 1716 UTC. The integrated water vapor retrieval (bottom row, third plot) shows an initial decrease from 0.58 cm at 0000 UTC to a relative minimum of 0.49 cm near the time of frontal passage at 1120 UTC, followed by a gradual increase within the post-cold frontal upslope flow to 0.63 cm at 2204 UTC. The radiometer rain sensor (not shown) did not indicate liquid precipitation accumulation during this period (Knupp, 2006).

A comparison of the radiometer sounding with the 1200 UTC February 16, 2001, radiosonde sounding at Denver (located 50 km southeast of Boulder) is shown in Figure 4.9. A deep temperature inversion extends from 400 to 900 AGL, high relative humidity exists below 500 m, and a tropopause height near 10 km is indicated in both the radiosonde and radiometer soundings. The retrieved fog/stratus water content maximum density of 0.14 g m^{-3} located near 300 m AGL is consistent with

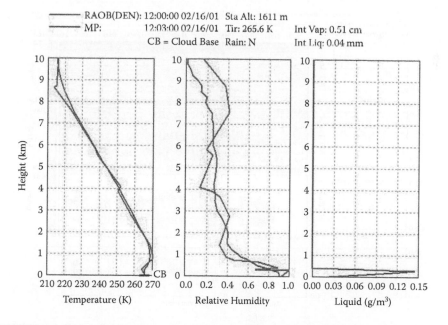

FIGURE 4.9 (See color insert.) Boulder retrieval (blue) and Denver radiosonde sounding (red) showing supercooled fog, an inversion at 1 km height and relative humidity saturation up to 300 m height at 1200 UTC on 16 Feb 01. The radiometric retrieval shows 0.04 mm integrated liquid water and 0.14 g/m^3 maximum liquid water density.

observations of fog (e.g., Gerber et al., 1999). Poor visibility and icing conditions during this upslope event led to major disruptions in surface and air transportation in the Denver area, including diversion of flights from Denver International Airport for 18 hours. Fog was not predicted by numerical forecasts that used the Denver radiosonde. However, variational assimilation of the Boulder radiometric retrievals led to accurate fog forecasts in the Boulder-Denver area (Vandenberghe and Ware, 2003).

Both the 0000 UTC February 16 Boulder radiometric retrieval (Figure 4.8) and Denver radiosonde sounding showed relative humidity less than 45% below 500 m height. During the following 11 hours, the 5-minute retrieved relative humidity increased somewhat prior to the rapid increase to >90% following the cold frontal passage after 1100 UTC. A similar trend was observed in Boulder prior to another event that produced super-cooled fog, freezing drizzle, and snow on March 4, 2003. Radiometric retrievals, weather radar, cloud radar, LIDAR, and tower observations during this weather event are described in Herzegh et al. (2003) and Rasmussen and Ikeda (2003). The high-resolution radiometer measurements in both cases suggest that trends in relative humidity can be used to predict the onset of fog. It would be better if raw radiometric brightness temperatures (rather than retrieved profiles) were directly assimilated into numerical weather models (Nehrkorn and Grassotti, 2003).

4.7.1.2 Snowfall

Radiometric retrievals during a light (trace accumulation) snowfall event in Boulder on December 23, 2002, are shown in Figure 4.10. Relatively dry snow that sublimated

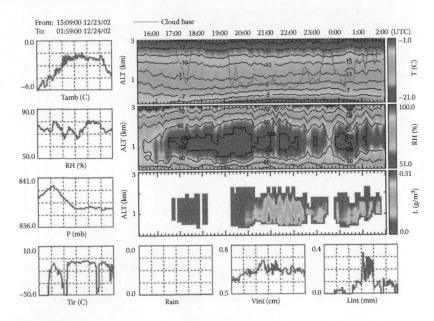

FIGURE 4.10 (See color insert.) Radiometric retrievals to 3 km height during snowfall showing relative humidity saturation near 1 km height and waves of "equivalent" liquid at 15 min intervals on 23 Dec 02 at Boulder. Equivalent integrated liquid and liquid density maxima of 0.33 mm and 0.31 g/m^3 respectively are seen just before 2100 UTC.

several minutes after touching the ground was visually observed at the radiometer site. Temperature profiles up to 3 km AGL (upper contour) are less than 0°C and show a gradual warming below 500 m AGL from 1600–2100 UTC, followed by a cooling trend. Relative humidity ranges from 90–100% between 0.5 and 2 km AGL during most of the observation period (middle contour). Equivalent liquid density variations with ~15 minute period and 0.1 g m^{-3} magnitude are seen from 500 m to 1.2 km AGL from 2050–2210 UTC (bottom contour plot). The liquid density retrieval is termed equivalent because the physical retrieval model is based on liquid emission only and does not include ice absorption and scattering, the effects of which would produce a very small positive bias for dry snow. If ice (and rain) absorption and scattering are included in the retrieval model, then liquid and ice retrievals can also be obtained (Li et al., 1997).

The Denver radiosonde sounding at 0000 UTC on December 24, 2002, and the Boulder radiometer sounding at 2354 UTC on December 23, 2002, are shown in Figure 4.11. Both soundings show a similar temperature profile. The radiometric temperature retrieval is 2–3 K warmer than the radiosonde sounding from 1–3 km AGL, and is ~8 K warmer at the tropopause. The radiosonde relative humidity measurement increases from 65% at the surface to 80% at 700 m height, and then remains above 75% up to 4.5 km height. The retrieved radiometer relative humidity increases from 80% at the surface to saturation at 0.9 km and remains saturated to 1.75 km.

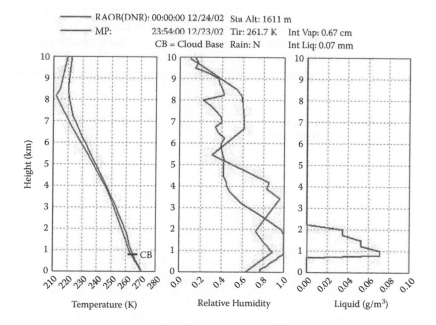

FIGURE 4.11 (See color insert.) Boulder retrieval (blue) and Denver radiosonde (red) during snowfall at Boulder on 23 Dec 02. The retrieval shows relative humidity saturation from 1 to 2 km height and 0.07 g/m³ equivalent liquid density near 1 km height. Tropopause height is seen in the radiosonde sounding and retrieval at 8.2 and 8.6 km height, respectively.

During the 11 hours prior to snowfall, relative humidity at 1 km AGL steadily increased from 50% to 100%. Since saturation is required for cloud and precipitation formation, the relative humidity profile trends could potentially improve local short-term cloud and precipitation forecasting.

4.7.1.3 Thermodynamics within Cloud Systems

The microwave radiometer (MPR) has potential for profiling the temperature, and hence buoyancy, within cloud systems, provided that retrievals are based on more sophisticated radiative transfer physics to account for scattering and emission by precipitation-sized water and ice. Currently, no other ground-based remote sensing technique is able to routinely probe the temperature within (precipitating) cloud systems over the 0–10 km AGL layer. May et al. (2003) recently reported that radio acoustic sounding system (RASS) measurements with a 50 MHz profiler did not successfully retrieve virtual temperatures within relatively vigorous convective systems due to the apparent turbulent breakup of acoustic wave fronts. A more robust passive microwave retrieval scheme could, in principle, utilize information from profiling radars that sample the reflectivity factor or Doppler spectra of the precipitation medium. The Doppler spectra can then be analyzed to retrieve precipitation-size distributions and particle types (Williams et al., 2007), which could provide input to the radiative transfer model.

4.7.1.4 Boundary Layer Processes

Because the MPR can continuously monitor the atmospheric boundary layer (ABL), rapid changes in nocturnal boundary layer (NBL) phenomena (e.g., gravity waves and bores) (see Knupp, 2006) can be monitored. Since the MPR resolution scales with height, such large gradients are often resolved quite well. Figure 4.12 shows a radiosonde-MPR sounding comparison for a strong nocturnal inversion in Huntsville.

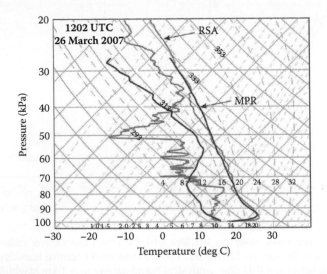

FIGURE 4.12   Nocturnal inversion soundings on 26 March 2007.

The temperature profile within the surface-based nocturnal inversion is duplicated quite well. The NBL and other statically stable layers are conducive to gravity wave propagation (Crook, 1996). The MPR combined with a 915 MHz profiler is able to resolve the presence of gravity waves in vertical motion, potential temperature, or humidity fluctuations, as well as the environment (e.g., Scorer parameter, which depends on vertical profiles of static stability and wind shear) that supports ducted gravity waves.

4.7.1.5   Severe Storms and Their Environment

Static stability, usually expressed in bulk form as the convective available potential energy (CAPE), is an important parameter in assessing severe storm potential. Likewise, analysis of convective inhibition is used to anticipate the timing of convective initiation. Rapidly changing vertical thermodynamic profiles are often produced locally through boundary layer heating or mesoscale vertical motion, and by horizontal temperature advection. Continuous monitoring of thermodynamic indices can therefore provide very timely information. Feltz et al. (2003) utilized the hyperspectral atmospheric emitted radiance interferometer (AERI) instrument to assess the evolution of CAP during a tornado outbreak. The University of Alabama in Huntsville (UAH) MPR is used to generate real-time soundings and CAPE estimates for the Huntsville, Alabama, National Weather Service office.

The MPR has been used by UAH researchers to determine the propagation modes of squall lines. Three different propagation modes have been identified, including the common density current mode (cold pool), which is very basic to squall line propagation and evolution (Bryan et al., 2006). In addition, the MPR has documented cases in which the squall line propagates as a bore and as a high-amplitude low-level ducted gravity wave. The latter has been considered previously in the form of a wave-CISK (Conditional instability of the Second Kind) mechanism over the troposphere. The role of the cold pool in the bore and gravity wave cases is not apparent, and may play a more minor role than is the case for density currents.

4.7.1.6   Quantitative Precipitation Forecasting (QPF)

QPF is at the top of this list because of the recognized need to assimilate water vapor fields into mesoscale forecast models, in order to improve accuracy of QPF (Weckwerth and Parsons, 2006). It is known that convective initiation in mesoscale models is very sensitive to relatively small three-dimensional (3D) variations in temperature and the mixing ratio. Initial numerical modeling shows that assimilation of detailed 3D water vapor fields produces a significant improvement in the model precipitation forecasts (Wulfmeyer et al., 2006).

4.7.1.7   Aviation Forecasting

The ability to detect super-cooled liquid is an obvious application of MPR measurements to aviation forecasting. This capability would make a potentially important contribution to the current icing potential (CIP) (Bernstein et al., 2005), which currently utilizes multiple data sources (e.g., surface, radar, satellite, forecast models, etc.).

4.7.1.8 Winter Weather Forecasting

MPR measurements have also been useful in winter precipitation forecasting, such as utilization of the lower atmospheric temperature/humidity profile to determine the relative probability of rain, freezing rain, and snow. This capability was used to explain an unexpected (but significant) sleet event around Huntsville, in which the entire lower atmosphere was just above 0°C, but dry conditions promoted a low wet-bulb temperature ($Tw < 0°C$), and hence cooling of raindrops and the initiation of raindrop freezing. Thus, profiles of Tw represent another useful application for real-time monitoring of winter weather.

Integrated measures of temperature, such as thickness, are extensively utilized to forecast winter precipitation type (e.g., Heppner, 1992). Radiometric measurements can be used quite effectively to monitor various thickness values (1000–925 hPa, 1000–850 hPa) since the integrated temperature measurements tend to reduce errors.

4.7.1.9 Severe Storms Forecasting

The MPR, used in conjunction with other profiling instruments such as Ultra high frequency (UHF) profiling radars, can provide continuous monitoring of CAPE (Feltz and Mecikalski, 2002), wind shear, and parameters derived from combinations of thermodynamic (Weckwerth, 2000) and wind profiles such as storm-relative helicity and energy helicity index (EHI), which are important ingredients in severe storm forecasts (Johns and Doswell, 1992).

REFERENCES

Bernstein, B.C., F. McDonough, M.K. Politovich, B.G. Brown, T.P. Ratvasky, D.R. Miller, C.A. Wolff, and G. Cunning. Current icing potential: Algorithm description and comparison with aircraft observations. *J. Appl. Meteorol.*, 44, 969–986, 2005.

Bryan, G.H., J.C. Knievel, and M.D. Parker. A multimodel assessment of RKW theory's relevance to squall-line characteristics. *Monthly Weather Rev.*, 134, 2772–2792, 2006.

Chong, W., and D. Lu. A universal regression retrieval method of the ground based microwave remote sensing of perceptible water vapour and path integrated cloud liquid water content. *Atmos. Res.*, 34, 309–322, 1994.

Cimini, D., E.R. Westwater, A.J. Gasiewski, M. Klein, V. Leuski, and J.C. Liljegren. Ground-based millimeter and submillimeter-wave observations of low vapor and liquid water contents. *IEEE Trans. Geosci. Remote Sensing*, 45, 2169–2180, 2007.

Cracknell, A.P., and C.A. Varotsos. The IPCC Fourth Assessment Report and the fiftieth anniversary of Sputnik. *Environ. Sci. Pollut. Res.*, 14, 384–387, 2007.

Crook, N.A. Sensitivity of moist convection forced by boundary layer processes to low-level thermodynamic fields. *Monthly Weather Rev.*, 124, 1767–1785, 1996.

Feltz, W.F., and J.R. Mecikalski. Monitoring high-temporal-resolution convective stability indices using the ground-based atmospheric emitted radiance interferometer (AERI) during the 3 May 1999 Oklahoma–Kansas tornado outbreak. *Weather Forecasting*, 17, 445–455, 2002.

Feltz, W.F., W.L. Smith, H.B. Howell, R.O. Knuteson, H. Woolf, and H.E. Revercomb. Near-continuous profiling of temperature, moisture, and atmospheric stability using the atmospheric emitted radiance interferometer (AERI). *J. Appl. Meteorol.*, 42, 584–597, 2003.

Franklin, J.N. *Matrix theory*. Prentice-Hall, Englewood Cliffs, NJ, 1968.

Franklin, J.N. Well-posed stochastic extension of ill-posed problems. *J. Math. Anal. Appl.*, 31, 682–716, 1970.

Gerber, H., Alfred R. Rodi, Aug 1999, Ground-Based FSSP and PVM. Measurements of Liquid Water Content, *Journal of Atmospheric and Oceanic Technology, Volume* 16, 1143–1149.

Goody, R.M., and Y.L. Yung. *Atmospheric radiation: Theoretical basis.* 2nd ed. Oxford University Press, Oxford, 1995.

Grant, E.H., T.J. Buchanan, and H.F. Cook. Dielectric behavior of water at microwave frequencies. *J. Chem. Phys.*, 26, 156–161, 1957.

Grody, C. Remote sensing of atmospheric water content from satellite using microwave radiometry. *IEEE Trans. Antennas Propagation*, 24, 155–162, 1976.

Heppner, P.O. Snow versus rain: Looking beyond the "magic" numbers. *Weather Forecasting*, 7, 683–691, 1992.

Herzegh, P., S. Landolt, and T. Schneider. The structure, evolution and cloud processes of a Colorado upslope storm as shown by profiling radiometer, radar, and tower data. Presented at Proceedings of 31st Conference on Radar Meteorology, Seattle, WA, 2003.

Ishimaru, A. *Wave propagation and scattering in random media.* Section 22-7, vol. II. Academic Press, New York, 1978.

Janssen, M.A. A new instrument for the determination of radiopath delay due to atmospheric water vapor. *IEEE Trans. Geosci. Remote Sensing*, 23, 455–490, 1985.

Johns, R.H., and C.A. Doswell. Severe local storms forecasting. *Weather Forecasting*, 7, 588–612, 1992.

Karmakar, P.K. *Microwave propagation and remote sensing.* CRC Press, Boca Raton, FL, 2011.

Karmakar, P.K., and S. Chattopadhyay. Radiosonde studies of microwave attenuation and water vapour—A review. *Indian J. Phys.*, 78B, 13–29, 2004.

Karmakar, P.K., S. Chattopadhyay, and A.K. Sen. Estimates of water vapour absorption over Calcutta at 22.235 GHz. *Int. J. Remote Sensing*, 20, 2637–2651, 1999.

Karmakar, P.K., M. Maiti, A.J.P. Calheiros, C.F. Angelis, L.A.T. Machado, and S.S. da Costa. Ground-based single-frequency microwave radiometric measurement of water vapour. *Int. J. Remote Sensing*, 32(23), 1–11, 2011a.

Karmakar, P.K., M. Maiti, S. Chattopadhyay, and M. Rahaman. Effect of water vapour and liquid water on microwave absorption spectra and its application. *Radio Sci. Bull.*, 303, 32–36, 2002.

Karmakar, P.K., M. Maiti, S. Sett, C.F. Angelis, and L.A.T. Machado. Radiometric estimation of water vapor content over Brazil. *Adv. Space Res.*, 48, 1506–1514, 2011b.

Karmakar, P.K., M. Rahaman, and A.K. Sen. Measurement of atmospheric water vapour content over a tropical location by dual frequency microwave radiometry. *Int. J. Remote Sensing*, 22(17), 3309–3322, 2001.

Karmakar, P.K., L. Sengupta, M. Maiti, and C.F. Angelis. Some of the atmospheric influences on microwave propagation through atmosphere. *Am. J. Sci. Ind. Res.*, 1(2), 350–358, 2010.

Karmakar, P.K., P.K. Tarafdar, S. Chattopadhya, and A.K. Sen. Studies on water vapour over a coastal region. *Indian J. Phys.*, 70B(3), 130–135, 1996.

Keihm, S.J., and K.A. Marsh. *Advanced algorithm and system development for Cassini radio science tropospheric calibration.* Progress report for the JPL Telecommunication and Data Acquisition Program, 1997.

Klein, M and A.J Gasiewski. Nadir sensitivity of passive millimeter and submillimeter wave channels to clear air and temperature and water vapor variations, *Journal of Geophysical Research*, 105, 17481–17511, 2000.

Knupp, K. Observational analysis of a gust front to bore to solitary wave transition within an evolving nocturnal boundary layer. *J. Atmos. Sci.*, 63, 2016–2035, 2006.

Knupp, K., R. Ware, D. Cimini, F. Vandenberghe, J. Vivekanandan, E. Westwater, and T. Coleman. Ground-based passive microwave profiling during dynamic weather conditions. 2012.

Li, L., J. Vivekanandan, C.H. Chan, and L. Tsang. Microwave radiometric technique to retrieve vapor, liquid and ice. *IEEE Trans. Geosci. Remote Sensing*, 35, 224–236, 1997.

Liebe, H.J. MPM—An atmospheric millimeter wave propagation model. *Int. J. Infrared Millimeter Waves*, 10, 631–650, 1989.

Liebe, H.J. An updated model of millimetre wave propagation in moist air. *Radio Science* 20, 5, 1069–1089, 1985.

May, P.T., C. Lucas, R. Lataitis, and I.M. Reid. On the use of 50-MHz RASS in thunderstorms. *J. Atmos. Oceanic Technol.*, 20, 936–943, 2003.

Nehrkorn, T., and C. Grassotti. Mesoscale variational assimilation of profiling radiometer data. Presented at 16th Conference on Numerical Weather Prediction, Seattle, WA, 2003.

Pandey Prem, C., B.S. Gohil, and T.A. Hariharan. A two frequency algorithm differential technique for retrieving precipitable water from satellite microwave radiometer (SAMIR-II) on board Bhaskara II. *IEEE Trans. Geosci. Remote Sensing*, 22, 647–655, 1984.

Rasmussen, R., and K. Ikeda. Radar observations of a freezing drizzle case in Colorado. Presented at Proceedings of 31st Conference on Radar Meteorology, Seattle, WA, 2003.

Resch, G.M. *Another look at the optimum frequencies for a water vapour radiometer*. TDA Progress Report 42-76. 1983. Available at http://ipnpr.jpl.nasa.gov/progress_report/42-76/76A.PDF (accessed May 13, 2011).

Rodgers, C.D. Retrieval of atmospheric temperature and composition from remote measurements of thermal radiation, *Review of Geophysics and Space Physics*, 14, 609–624, 1976.

Rodgers, C.D. *Statistical principle of inversion theory, in Inversion Methods in Atmospheric Remote Sensing*, (A. Deepak ed), Academic Press, New York, 117–138, 1977.

Rosenkranz, P.W. Interference coefficients for overlapping oxygen lines in air. *J. Spectrosc. Radiative Transfer*, 39, 287–297, 1988.

Sen, A.K., P.K. Karmakar, S. Dev Barman, M.K. Das Gupta, O.P.N. Calla, and S.S. Rana. Tropospheric water vapour modelling over a tropical location by radiometric study. *Indian J. Radio Space Phys.*, 20, 347–350, 1991.

Sen, A.K., P.K. Karmakar, A. Maitra, A.K. Devgupta, M.K. Dasgupta, O.P.N. Calla, and S.S. Rana. Radiometric studies of clear air attenuation and atmospheric water vapour at 22.235 GHz over Calcutta (lat. 22°N, long. 88°E). *Atmos. Environ.*, 24A(7), 1909–1913, 1990.

Simpson, P.M., E.C. Brand, and C.L. *Wrench. Liquid water path algorithm development and accuracy. Microwave radiometer measurements at Chilbolton*. Radio Communications Research Unit, CLRC—Rutherford Appleton Laboratory, Chilton, DIDCOT, Oxon, UK, 2002.

Solheim, F., J.R. Godwin, E.R. Westwater, H. Yong, S.J. Keihm, K. Marsh, and R. Ware. Radiometric profiling of temperature, water vapour and cloud liquid water using various methods. *Radio Sci.*, 33, 393–404, 1998.

Tzanis, C., and C.A. Varotsos. Tropospheric aerosol forcing of climate: A case study for the greater area of Greece. *Int. J. Remote Sensing*, 29, 2507–2517, 2008.

Ulaby, F.T., R.K. Moore, and A.K. Fung. *Microwave remote sensing—Active and passive*. Vol. I. Addison-Wesley, Reading, MA, 1981.

Ulaby, F.T., R.K. Moore, and A.K. Fung. *Microwave remote sensing—Active and passive*. Vol. III. Artech House, Norwood, MA, 1986.

Vandenberghe, F., and R. Ware. 4-dimensional variational assimilation of ground-based microwave observations during a winter fog event. Presented at International Workshop on GPS Meteorology, Tsukuba, Japan, 2003.

Weckwerth, T.M. The effect of small-scale moisture variability on thunderstorm initiation. *Monthly Weather Rev.*, 128, 4017–4030, 2000.

Weckwerth, T.M., and D.B. Parsons. A review of convection initiation and motivation for IHOP_2002. *Monthly Weather Rev.*, 134, 5–22, 2006.

Westwater, E.R. Ground-based determination of low altitude temperature profiles by microwaves. *Monthly Weather Rev.*, 100(1), 15–28, 1972.

Westwater, E.R. The accuracy of water vapor and cloud liquid determination by dual-frequency ground-based microwave radiometry. *Radio Sci.*, 13, 677–685, 1978.

Westwater, E.R. Initial results from the 2004 North Slope of Alaska Arctic Winter Radiometric Experiment. Presented at Proceedings of IGARSS'04, Anchorage, AK, September 20–24, 2004.

Westwater, E.R., S. Crewell, C. Mätzler, and D. Cimini. Principles of surface based microwave and millimeter wave radiometric remote sensing of the troposphere. *Quaderni Della Societa Italiana Di Electromagnetsimo*, 1, 3, 2005.

Westwater, E.R., and M.T. Decker. Application of statistical inversion to ground based microwave remote sensing of temperature and water vapour profiles. In *Inversion methods in atmospheric remote sounding*, ed. A. Deepak. Academic Press, New York, pp. 395–428, 1977.

Westwater, E.R., and N.C. Grody. Combined surface and satellite based microwave temperature profile retrieval. *J. Appl. Meteorol.*, 19, 1438–1444, 1980.

Westwater, E.R., and F.O. Guiraud. Ground-based microwave radiometric retrieval of precipitable water vapor in the presence of clouds with high liquid content. *Radio Sci.*, 13, 947–957, 1980.

Westwater, E.R., and J.B. Snider. Ground based radiometric observation of atmospheric emission and attenuation at 20.6, 31.65, and 90.0 GHz: A comparison of measurements and theory. *IEEE Trans. Antennas Propagation*, 38(10), 1569–1580, 1990.

Williams, C.R., A.B. White, K.S. Gage, and F.M. Ralph. Vertical structure of precipitation and related microphysics observed by NOAA profilers and TRMM during NAME 2004. *J. Climate*, 20, 1693–1712, 2007.

Wulfmeyer, V., H.S. Bauer, M. Grzeschik, A. Behrendt, F. Vandenberghe, E.V. Browell, S. Ismail, and R.A. Ferrare. Four-dimensional variational assimilation of water vapor differential absorption LIDAR data. The first case study within IHOP_2002. *Monthly Weather Rev.*, 134, 209–230, 2006.

5 Ground-Based Radiometric Sensing of Thermodynamic Variables in the Polar Regions

5.1 INTRODUCTION

In principle, radiometric data can be utilized to infer the height profile of water vapor density, and temperature as well, using the statistical inversion techniques discussed in Chapter 4. The selection of frequency demands the following important criteria: (1) the radiometric temperature at selected frequencies should be strongly sensitive to water vapor density and only weakly sensitive to other atmospheric variables, and (2) the water vapor weighting function at the selected frequencies should have height profiles sufficiently different to minimize redundancy. We know that the water vapor weighting function $W\rho(f, h)$ is defined through the standard form (Ulaby et al., 1986) as

$$T_{DN} = \int_0^\infty W_\rho(f,h)\rho_V(h)dh \tag{5.1}$$

where

$$W_\rho(f,h) = k_f(h)\frac{T(h)}{\rho_V(h)}e^{-\tau(0,h)}\text{K/km.g.m}^{-3} \tag{5.2}$$

Here $k_f(h)$ accounts for the absorption coefficient of oxygen and water vapor. The water vapor absorption coefficient dominates the situation at or near the peak of the spectrum. This is also true for 183 GHz, a strong water vapor resonance line. If we look at the plot (Figure 5.1) of attenuation (dB), as reported by Smith (1982), for typical atmosphere, we see that at $f = 22.235$ GHz, k_{water} is larger than k_{oxygen} by a factor of 20 ($\rho_0 = 7.5\ gm^{-3}$). In a dry climate, for ($\rho_0 = 1.0\ gm^{-3}$), the ratio comes down to 3 only, and hence the total absorption coefficient is not as strongly dominated by water vapor.

Traditionally, ground-based radiometric measurements employ two or more channels located in the spectral region of water vapor absorption. One is at 23.835 GHz, as this is independent of pressure broadening, and one channel in the window region

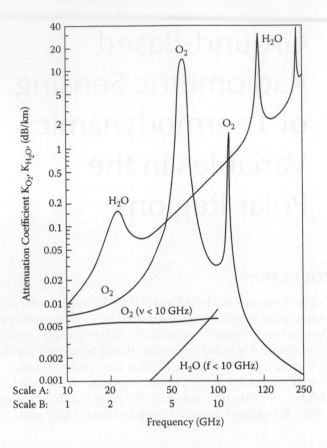

FIGURE 5.1 Oxygen and water vapor spectra at 290 K and 7.5 g/m³ of 1013 mb for the dry atmosphere and 102.5 mb for the moist atmosphere.

near 30 GHz, where liquid water absorption dominates during cloudy conditions. But in polar regions, including arid regions, high latitudes, deserts, or above the atmospheric boundary layer, the water vapor measurement accuracy of a few tenth of a millimeter is required to monitor changes in humidity. During the cold Arctic winter, the amount of precipitable water vapor is often less than 3 mm, and clouds with liquid water less than 50 g-m⁻² are common (Maria et al., 2007). In fact, the Arctic region is characterized by extremely dry conditions; during winter, precipitable water vapor (PWV) and liquid water path (LWP) are usually below 5 mm (Racette et al., 2005) and 0.2 mm (Shupe et al., 2005), respectively. It follows that the accuracy of existing instrumentation is limiting the development of theory and modeling of the Arctic radiative processes. On the other hand, millimeter (mm) and submillimeter wavelengths offer a powerful tool to increase the sensitivity during Arctic conditions (Cimini et al., 2007b).

Since the discovery of the ozone hole in the mid-1980s over Antarctica and the Montreal Protocol phasing out the production of freons, the main compounds that destroy the ozone layer, the international scientific community has focused on the understanding of the photochemical processes acting in the middle atmosphere. Chemistry-climate interactions are far from being known and could modify even more the evolution of the ozone layer by favoring more stable and colder polar vortices, thus increasing the probability of occurrence of polar stratospheric clouds (PSCs), consequently increasing chlorine activation and slowing down the recovery of the ozone layer (WMO, 2006). Water vapor (H_2O) plays a key role in the earth climate system since it is the main greenhouse gas emitting and absorbing in the infrared domain. Its variability in both the troposphere and the stratosphere is still an enigma, or at least is still under discussion (Scherer et al., 2008). At high latitudes, and more precisely over Antarctica in winter, the stratospheric polar vortex develops within a cold and dry atmosphere where physicochemical processes are globally well known (Brasseur et al., 1999): the presence of PSC, dehydration, halogen activation (chlorine and bromine compounds) via heterogeneous chemical reactions and solar illumination, sedimentation, descent within the core of the vortex, and de-nitrification. All of these processes produce ozone loss. Nevertheless, great uncertainties (differences between measurements and model) pertain or are linked to the quantification of the ozone loss rate, de-nitrification rate, descent rate, and dehydration rate within the polar vortex. Several campaigns have been organized by the European community in the northern hemisphere to understand the destruction mechanisms within polar vortices (Pyle et al., 2004). These campaigns have not taken place in the southern hemisphere so far. The Antarctica Microwave Stratospheric and Tropospheric Radiometers (HAMSTRAD) program aims to develop two ground-based microwave radiometers to sound tropospheric and stratospheric H_2O above Dome C (Concordia Station), Antarctica (75°S, 123°E) over a long time period. HAMSTRAD-Tropo is a 183 GHz radiometer for measuring tropospheric H_2O.

The HAMSTRAD radiometers will complement two other instruments: an ultraviolet-visible Système d'Analyse par Observation Zénithale instrument (Pommereau and Goutail, 1988) for the measurements of ozone and nitrogen dioxide columns and detection of PSCs, and an ultraviolet (UV) photometer (Bais et al., 2001) to measure UVB radiation in the domain 280–320 nm at the surface. These instruments intend to give some elements of answer to important questions relative to the evolution of the ozone layer, to the constituents linked to its destruction, and to dynamical processes coupled to the polar vortex evolution.

In these conditions the dual-channel radiometer operates at the limit of its capabilities with a very low signal-to-noise ratio. Several attempts have been made from time to time to find out the origin of uncertainties in vapor retrieval (Crewell and Lohnert, 2003). However, the uncertainties in the liquid water retrieval can be attributed to the modeling of the dry opacity term and to the cloud liquid absorption coefficients, in addition to calibration issues and the effect of measurement noise. However, it is clear that it has not been possible to obtain accurate retrieval of water vapor density from radiometric observation at or near the weak resonance line 22.235 GHz, especially in dry climates. We attributed this lack of success to the weakness of the 22.235 GHz line, as a result of which the weighting functions for water are approximately uniform with altitude (Figure 5.2a,b). This is equivalent to saying that radiometric observations at these frequencies in polar

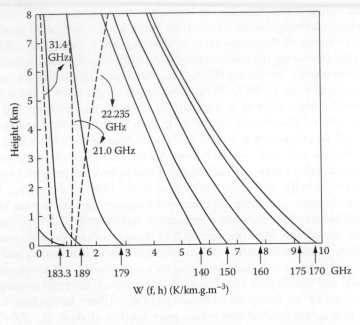

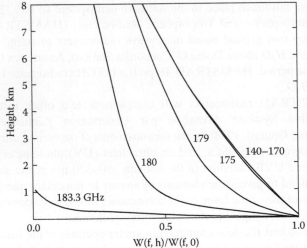

FIGURE 5.2 (a) Water vapor weighting function around 22.234 and 183.31 GHz. (b) Normalized water vapor weighting function for frequencies ranging from 140–183.3 GHz.

regions have a very poor vertical resolution. For the sake of clarity, a plot (Cimini et al., 2007b) of atmospheric opacity for standard Arctic conditions (Figure 5.3) is presented.

The increased sensitivity at 183.31 GHz to water vapor can help improve water vapor retrievals during dry Arctic winter. It is worthwhile to mention that the 183 GHz line is about 50 times more sensitive to changes in precipitable water vapor

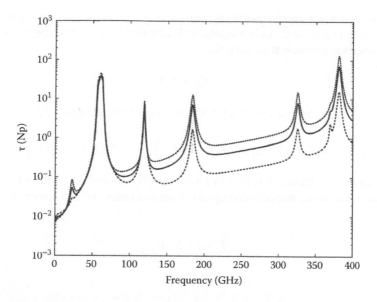

FIGURE 5.3 Atmospheric opacity for standard Arctic conditions with 1, 5, and 10 mm PWV (shown with dash-dotted, solid, and dashed lines, respectively). Vertical dotted lines indicate the spectral location of MWR, MWRP, and GSR channels.

and about 10 times more sensitive to liquid water (Racette et al., 2005) than that at the 22 GHz absorption line. In dry conditions, 183 GHz brightness temperature changes by more than 20 K (Cimini et al. 2010). For each millimeter change in water vapor column, an instrument with a 1 K radiometric measurement precision can detect 0.05 mm change in the vapor column. But this 1 K precision needed at 183 GHz is a routine radiometer design goal. Nevertheless, at this wavelength, the front-end losses, effects of radome reflection, and complexity of incorporating stable calibration loads in the radiometer, present significant design challenges. However, the recent advances make the development of a 183 GHz radiometer more practical today.

5.2 THEORETICAL BACKGROUND

We already know that the radiative transfer equation can be written as

$$B_f\left(T_b\right) = B_f\left(T_c\right)e^{-\tau_V} + B_f\left(T_m\right)(1 - e^{-\tau_V}) \tag{5.3}$$

where B_f is the Planck function, and τ is the opacity at a frequency f. Now considering the clear sky, i.e., without cloud, we write

$$\tau_{tot} = \tau_{dry} + \tau_{vapour} = \tau_{dry} + \int_0^\infty \alpha_V(h)\rho_V(h)dh \tag{5.4}$$

where $\alpha_V(h)$ is water vapor absorption coefficient at a particular height of choice.

Assuming $\alpha_V(h)$ varies only slightly with altitude, let us introduce the average mass absorption coefficient k_V, such that

$$\tau = \tau_{dry} + k_V V \tag{5.5}$$

Now in the case of low opacity ($\tau \leq 0.5 \, np$), we write from Equation 5.3:

$$B_f(T_b) = B_f(T_C) + \tau_d \left[B_f(T_m) - B_f(T_C) \right] + k_V \left[B_f(T_m) - B_f(T_C) \right] V \tag{5.6}$$

Expanding the Planck function in terms of hf/kT (h and k are the Planck and Boltzmann constants, respectively) up to the second order, we can rewrite Equation 5.6 as

$$T_b = I + S \cdot V \tag{5.7}$$

where

$$I = T_C + \tau_d(T_m - T_C) + (1 - \tau_d)\left(\frac{hf}{k}\right)^2 \cdot \frac{1}{12T_C} \tag{5.8}$$

and

$$S = k_V(T_m - T_C) - \left(\frac{hf}{k}\right)^2 \cdot \frac{1}{12T_C} \tag{5.9}$$

5.3 WEIGHTING FUNCTION ANALYSIS

As reported by Cimini et al. (2007a), the Arctic Winter Radiometric Experiment (Westwater et al., 2006) was held at the Atmospheric Radiation Measurement (ARM) Program's North Slope of Alaska (NSA) site near Barrow, Alaska, from March 9 to April 9, 2004. During this experiment, the 25-channel ground-based scanning radiometer (GSR) was first deployed. This instrument includes 5 radiometers providing 12 channels in the low-frequency wing of the 60 GHz oxygen complex (50.2, 50.3, 51.76, 52.625, 53.29, 53.845, 54.4, 54.95, 55.52, 56.025, 56.215, and 56.325 GHz), 2 channels at 89 GHz (horizontal (H) and vertical (V) polarizations), 7 channels distributed around the 183.31 GHz water vapor absorption line (183.31 ± 0.55, ± 1, ± 3.05, ± 4.7, ± 7, ± 12, ± 16 GHz), 2 polarized channels at 340 GHz (H and V), and 3 channels around the 380.2 GHz water vapor line (380.197 ± 4, ± 9, ± 17 GHz). The channels were selected to provide simultaneous retrievals of precipitable water vapor (PWV), liquid water path (LWP), and low-resolution temperature and humidity profiles.

The incremental weighting functions (Westwater, 1993) give an estimate of the sensitivity of a particular channel to changes of a given atmospheric variable (e.g., temperature, water vapor density, and liquid water content), and thus indicate the ability to retrieve that particular parameter from passive observations. In Figure 5.4, weighting function profiles for selected millimeter-wave radiometer (MWR), millimeter-wave radiometer profiler (MWRP), and ground-based scanning radiometer (GSR) channels as computed using the Arctic atmospheres were introduced (PWV = 1, 5, 10 mm). Figure 5.4(a) shows the water vapor weighting function, WFρ, for the two MWR channels at 23.8 and 31.4 GHz, the MWRP channel at 22.235 GHz, and the GSR channel at 89 GHz. Note that the MWRP channel is located at the center of the water vapor line, thus corresponding to the maximum

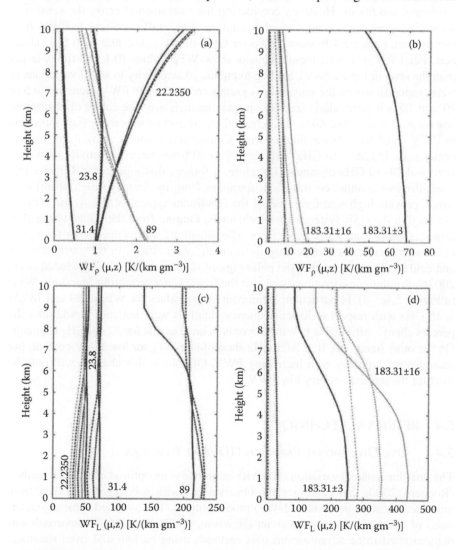

FIGURE 5.4 (a, b) Water vapor and (c, d) liquid water weighting functions. WF is a function of height z for selected MWRP and GSR channels. WFs were computed with a standard Arctic atmosphere, setting PWV to 1, 5, and 10 mm (shown with solid, dotted, and dashed lines, respectively). Zenith observations are considered in this plot (air mass $\mu = 1$).

absorption in the 20–30 GHz range, whereas 89 GHz corresponds to the lowest absorption among the GSR channels. WFρ values at 23.8 and 31.4 GHz show little variation with height, which makes these two channels optimal for estimates of integrated contents. Conversely, WFρ values at 22.235 and 89 GHz show vertical structure with inverse trends. In any case, these four channels exhibit sensitivity to water vapor of the same order of magnitude, with almost no dependence on water vapor content. This feature is kept for higher PWV contents (not shown), which makes low-frequency channels eligible for water vapor observations in any environmental conditions. However, considering the instrumental error, the sensitivity shown in Figure 5.4(a) is not enough to capture very small variations in PWV. As a comparison, Figure 5.4(b) shows WFρ for GSR 183.31 ± 3.05 and ± 16 GHz channels. For PWV = 1 mm, these channels show WFρ values 10 to 60 times larger than the ones in Figure 5.4(a), leading to enhanced sensitivity to small variations in water vapor. However, the sensitivity is greatly reduced when PWV increases to 5 or 10 mm. This is particularly true for opaque channels near the center of absorption lines, e.g., 183.31 ± 3.05 GHz. This channel has almost no sensitivity (i.e., saturates) to PWV equal to or larger than 10 mm. Conversely, channels away from the line center (e.g., 183.31 ± 16 GHz) still show 5 to 10 times larger sensitivity than conventional 20–30 GHz channels. Therefore, it follows that a proper millimeter and submillimeter channel combination, spanning from moderate to high absorption, would provide high sensitivity for all the conditions typical of the Arctic. This is obtained in the GSR system with 10 channels, ranging from the wings to the near center of two strong water vapor lines. The sensitivity makes the higher-frequency radiometers particularly appealing for accurate observations in the extremely dry and cold conditions typical of the polar regions (Cimini et al., 2007; Cadeddu et al., 2007b). Similar considerations apply for the liquid water weighting functions, WF_L, in Figure 5.4(c, d). In particular, significant larger values for WF_L at 89 and 183.31 ± 16 GHz with respect to lower-frequency channels were noticed. In addition, the percent change attributable to water vapor is as large as that for 20–30 GHz channels. On the other hand, 183.31 ± 3.05 GHz shows large WF_L for low PWV content, but saturates very quickly with increasing PWV. Of course, this channel will only be used for the retrieval of very low PWV.

5.4   RETRIEVAL TECHNIQUE

5.4.1   One-Dimensional Variation (1DVAR) Technique

The one-dimensional variation (1DVAR) technique is an optimal estimation method (Rodgers, 2000) that combines the observations with a background taken from numerical weather prediction (NWP) model outputs. The assumed error characteristics of both are taken into account (Hewison, 2007). The 1DVAR approach was demonstrated to be advantageous over methods using background from statistical climatology (Cimini et al., 2006). In fact, as background information, 1DVAR uses a forecast state vector, which is usually more representative of the actual state than a climatological mean. A comparative analysis between a variety of retrieval methods applied to ground-based observations from conventional microwave radiometers (such as MWRP) indicated that the 1DVAR technique outperforms the other considered methods, these being based on various kinds of multiple regression and neural networks. Thus, it seemed convenient to couple the sensitivity of millimeter-wave radiometry with the advantages of the 1DVAR technique for the retrieval of temperature and humidity profiles in the Arctic (Cimini et al., 2010). For this technique the standard notation is as used by Ide et al. (1997), and B and R indicate the error covariance matrices of the background and observation vector $\mathbf{y}$, respectively. In addition, the forward model operator (i.e., radiative transfer model) with $F(\mathbf{x})$ was used. Thus, the technique adjusts the state vector $\mathbf{x}$ from the background state vector X_b to minimize the following cost function:

$$J = [y - F(X)]^T R^{-1} [y - F(X) + (X - X_b)^T B^{-1} [X - X_b]]$$
(5.10)

Here T and -1 represent the matrix transpose and inverse. The radiometric noise, representativeness, and forward model errors all contribute to the observation error covariance R. The minimization is achieved using the Levemberg-Marquardt method; this method was found to improve the convergence rate with respect to the classic Gauss-Newton method (Hewison, 2007) by introducing a factor γ that is adjusted after each iteration, depending on how the cost function J has changed; thus, calling K the Jacobian matrix of the observation vector with respect to the state vector, the solution

$$X_{i+1} = X_i + [(1+\gamma)B^{-1} + K_i^T R^{-1} K_i]^{-1} \cdot [K_i^T R^{-1}(y - F(X_i) - B^{-1}(X_i - X_B)]$$
(5.11)

is iterated until the following convergence criterion is satisfied:

$$[F(X_{i+1}) - F(X_i)]^T S^{-1} [F(X_{i+1}) - F(X_i)] \ll N \text{ (observations)}$$
(5.12)

Here

$$S = R(R + K_i B K_i^T)^{-1} R$$
(5.13)

and N(obs) indicates the number of observations (i.e., the dimension of $\mathbf{y}$).

The GSR was first deployed during the Water Vapor Intensive Operational Period (WVIOP) (March–April 2004), and later during the Radiative Heating in Underexplored Bands Campaign (RHUBC) (February–March 2007), both held at the ARM Program's NSA site in Barrow, Alaska (Ackerman and Stokes, 2003).

The state vectors used by Cimini et al. (2010) are profiles of temperature and total water (i.e., total of specific humidity and condensed water content) (Deblonde and English, 2003). The choice of total water has the advantages of reducing the dimension of the state vector, enforcing an implicit correlation between humidity and condensed water, and including a super-saturation constraint. Moreover, the introduction of a natural logarithm of total water creates error characteristics that are more closely Gaussian and prevents unphysical retrieval of negative humidity. The background error covariance matrices B for both temperature and humidity profiles may be computed from a set of simultaneous and co-located forecast-RAOB data (in both clear and cloudy conditions). This calculation of B inherently includes forecast errors as well as instrumental and representative errors from the radiosondes. The radiosonde instrumental error is assumed to be negligible compared with the representative error, which consists of the error associated with the representation of volume data (model) with point measurements (radiosondes). The B matrix including these terms seems appropriate for the radiometric retrieval minimization; since the grid cell of the NWP model is much larger than the radiometer observation volume, the latter can be assumed as a point measurement compared with the model cell, similar to radiosondes. It may be assumed that the B matrix estimated for humidity is valid for control variable total water, since no information on the background cloud water error covariance was available. This assumption is strictly valid during clear sky conditions only, while it underestimates the background error in cloudy conditions. The implications are that under cloudy conditions, humidity retrieval would rely more on the background and less on measurements than would be possible by adopting a B matrix that includes both humidity and liquid water errors. However, considering the infrequent and optically thin cloudy conditions encountered during RHUBC, it is understood that this assumption does not affect results significantly.

The observation vector is defined as the vector of T_B measured by GSR at a number of elevation angles, plus the surface temperature and humidity given by the sensors mounted on the lowest level (2 m) of the meteorological tower. The observation error covariance matrix R may be estimated using the GSR data taken from the WVIOP, following the approach by Hewison (2007). The forward model $F(X)$ is provided by the National Oceanic and Atmospheric Administration (NOAA) microwave radiative transfer code (Schroeder and Westwater, 1991), which also provides the weighting functions that were used to compute the Jacobians K with respect to temperature, humidity, and liquid water. The typical errors with respect to band-averaged T_B are within 0.1 K and were accounted for in the forward modeling component of the observation error.

The 1DVAR retrieval technique and the settings described in the previous section were applied to GSR data collected during the 3-week duration of RHUBC. These observations were found to be consistent with simultaneous and co-located observations from the other two independent 183 GHz radiometers and with simulations obtained from RAOBs (Cimini et al., 2009), generally within the expected accuracy. A typical example of temperature and humidity retrieved profiles is illustrated in Figure 5.5, simultaneous data obtained from the GSR brightness temperatures observed within 10 minutes of the radiosonde launch time. As a comparison, the background NWP profiles that were used as a first guess were shown, as well as the in situ observations from the radiosonde.

Concerning the temperature profile, it is noted that in this case, the NWP forecast is in good agreement with the RAOB, particularly in the atmospheric layer from 0.5 to 3.0 km. Conversely, in the upper part of the vertical domain (3–5 km), the NWP forecast shows about a 1–2 K bias with respect to the RAOB, while in the very first layer (0–0.5 km), it differs from RAOB by more than 10 K. Conversely, the 1DVAR retrieval agrees better with the RAOB in the lowest levels, while for the upper levels, the retrieved temperature tends to lie over the NWP background.

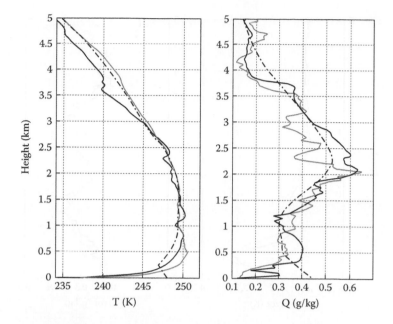

FIGURE 5.5 Typical example of (in gray) 1DVAR retrievals compared with the background from (black dashed) NWP and (black solid) RAOB temperature and humidity profiles (RAOB launched at 1624 on March 11, 2007).

As for the humidity, again, it is noted that the NWP forecast captures well the vertical structure, although with lower resolution, except for the first 500 m, where the 1DVAR retrieval shows a much better agreement with the RAOB. However, considering the NWP error analysis in Figure 5.6 and that the humidity Jacobians of 183 GHz channels are relatively smooth with respect to height, it is reasonable that the retrieved humidity departs from the background throughout the vertical domain.

5.5 WATER VAPOR OVER ANTARCTICA

Philippe Ricaud et al. (2010) reported that the geographical situation of Dome C (high altitude and dry air) is particularly well adapted to the setting of other instruments for monitoring the stratosphere and troposphere, for instance, the microwave radiometers. The altitude of the Dome C site is associated with a weak amount of water vapor in the troposphere, and the very low temperatures encountered in the lowermost altitude layers (one of the driest and coldest sites around the world) favor the setting up of microwave radiometers at high frequency and with a much better sensitivity (weak integration time) than sites located at sea level, in order to detect both stratospheric and tropospheric water vapor. The high latitude of the site might also help in validating the analyses of the European Centre for Medium-Range Weather Forecast (ECMWF). It is rather challenging to detect H_2O with ground-based microwave radiometers installed at very cold and dry areas, as they can be encountered at high elevation or high latitudes. The generally used $6_{16} - 5_{23}$ transition

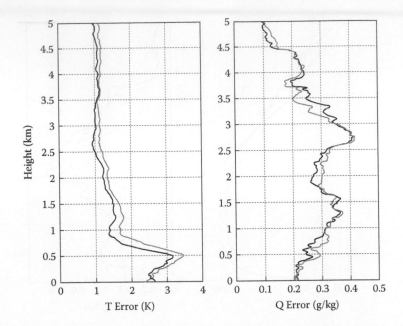

FIGURE 5.6 Diagonal terms (square root) of the background error covariance matrices B of (left) temperature and (right) humidity profiles from (gray) ECMWF and (black) National Centers for Environmental prediction (NCEP) global forecast near the ARM NSA site in Barrow, as computed from 77 forecast-RAOB matchups during the WVIOP 2004.

line at 22.235 GHz (Nedoluha et al., 1995; Forkman et al., 2003) has the main advantage of being detectable in the majority of sites around the world, but has the main drawback of being rather weak for measurement in very dry and cold conditions. It should be mentioned that water vapor distributions from the tropics to the driest places on earth (Antarctica) can vary over more than two orders of magnitude (e.g., IWV up to 70 kg · m^{-2} in the tropics and 0.15 kg · m^{-2} at Dome C). Recently, some ground-based microwave radiometers have been set up to detect the $3_{13} - 2_{20}$ transition line at 183.310 GHz (Westwater et al., 2006) to estimate the vertical distribution of tropospheric humidity. Within the Arctic Winter Experiment held at the Atmospheric Radiation Measurement Program's North Slope of Alaska site near Barrow in 2004 (Cimini et al., 2007b), the ground-based scanning radiometric measurement at 183 GHz was deployed along with 22 GHz radiometric studies supported by radio-sounding launches. Global results show that in extremely dry conditions, the precipitable water vapor (ranging from 1–2 mm) can be estimated within 5% uncertainty. It was also demonstrated that the 183 GHz radiometer contains channels that receive a significant response from the upper tropospheric region (Mattioli et al., 2008), although the following were observed: (1) some daytime dry biases were detected in the radio soundings (Mattioli et al., 2007), and (2) the stratospheric contribution might be nonnegligible when integrated water vapor (IWV) is less than 1 mm (Mattioli et al., 2008).

A statistical approach is used to calculate the profiles from the brightness temperatures measured by the radiometer. The retrieval algorithms are based on a few thousand radio soundings from the area where the instrument is deployed (namely, PdM and Dome C), which provide a set of temperature and humidity data points as a function of altitude. From radiative transfer calculations using theoretical spectral models like in Liebe (1989) and Rosenkranz (1998), a corresponding set of brightness temperatures measured on the surface of the two stations and at the radiometer frequencies is derived. The linear or quadratic regression analyses were applied to solve the inverse problem, namely, to estimate humidity and temperature profiles from the brightness temperature sets. During radiometer operation, the statistical fit coefficients provide an online determination of the tropospheric profiles from the measured brightness temperatures. Consequently, the *a priori* information from the radio soundings is introduced into the retrievals.

Figure 5.7 (Ricaud et al., 2010) shows the very first set of water vapor measurements from HAMSTRAD-Tropo covering the month of February 2008, from the 8th

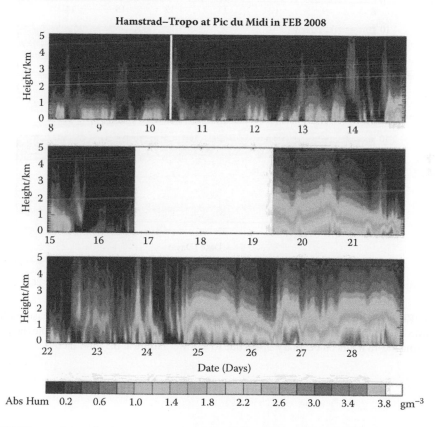

FIGURE 5.7 (See color insert.)   Temporal evolution of the absolute humidity as measured by HAMSTRAD-Tropo above the altitude of PdM (altitude = 0) in February 2008. Note the gap covering the 48 hours on February 17–18 due to a power failure at the PdM facility.

to the 28th. Note the gap covering the 48 hours on February 17–18 due to a power failure at the PdM facility. Figures 5.8 and 5.9 (Ricaud et al., 2010) also show, over the same period, the temporal evolution of temperature vertical profiles and of the temperature anomaly (difference between temperature and averaged temperature fields), respectively, as measured by HAMSTRAD-Tropo.

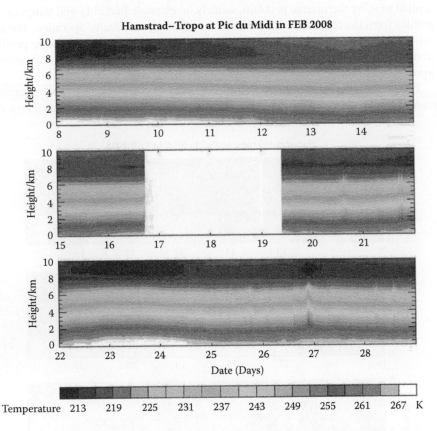

FIGURE 5.8 (See color insert.) Temporal evolution of the temperature as measured by HAMSTRAD-Tropo above the altitude of PdM (altitude = 0) in February 2008. Note the gap covering the 48 hours on February 17–18 due to a power failure at the PdM facility.

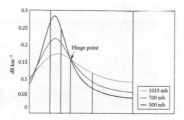

FIGURE 4.1 Representative plot of showing the occurrence of pressure independent frequency 23.834 GHz.

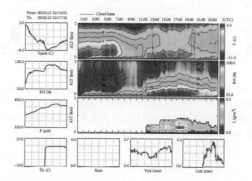

FIGURE 4.8 Radiometric retrievals of a supercooled fog event associated with upslope flow at Boulder on 16 Feb 2001. Poor visibility and icy conditions during this upslope event led to major transportation disruptions including diversion of international flights from Denver for 18 hours. Time series of temperature (upper contour), humidity (middle contour), and cloud liquid density (lower contour) profiles are shown along with surface temperature (left row, top plot), humidity (left row, second plot), and pressure (left row, third plot); zenith infrared temperature (left column, bottom plot; rain flag (lower row, second plot); and integrated water vapor (lower row, third plot) and integrated liquid water retrievals (lower row, fourth plot).

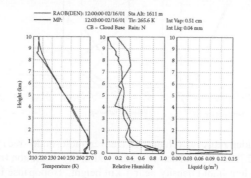

FIGURE 4.9 Boulder retrieval (blue) and Denver radiosonde sounding (red) showing supercooled fog, an inversion at 1 km height and relative humidity saturation up to 300 m height at 1200 UTC on 16 Feb 01. The radiometric retrieval shows 0.04 mm integrated liquid water and 0.14 g/m^3 maximum liquid water density.

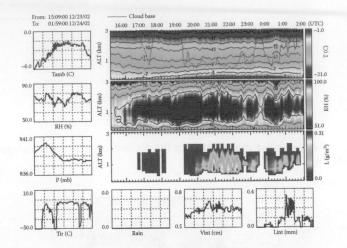

FIGURE 4.10 Radiometric retrievals to 3 km height during snowfall showing relative humidity saturation near 1 km height and waves of "equivalent" liquid at 15 min intervals on 23 Dec 02 at Boulder. Equivalent integrated liquid and liquid density maxima of 0.33 mm and 0.31 g/m³ respectively are seen just before 2100 UTC.

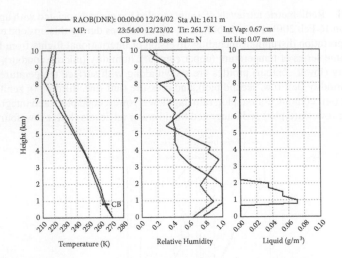

FIGURE 4.11 Boulder retrieval (blue) and Denver radiosonde (red) during snowfall at Boulder on 23 Dec 02. The retrieval shows relative humidity saturation from 1 to 2 km height and 0.07 g/m³ equivalent liquid density near 1 km height. Tropopause height is seen in the radiosonde sounding and retrieval at 8.2 and 8.6 km height, respectively.

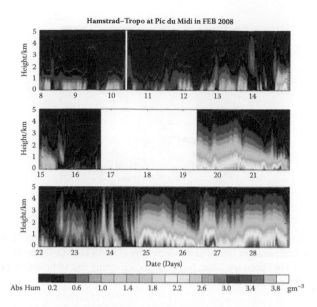

FIGURE 5.7 Temporal evolution of the absolute humidity as measured by HAMSTRAD-Tropo above the altitude of PdM (altitude = 0) in February 2008. Note the gap covering the 48 hours on February 17–18 due to a power failure at the PdM facility.

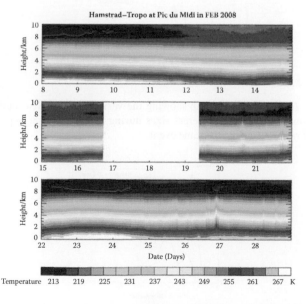

FIGURE 5.8 Temporal evolution of the temperature as measured by HAMSTRAD-Tropo above the altitude of PdM (altitude = 0) in February 2008. Note the gap covering the 48 hours on February 17–18 due to a power failure at the PdM facility.

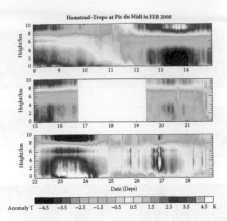

Anomaly T −4.5 −3.5 −2.5 −1.5 −0.5 0.5 1.5 2.5 3.5 4.5 K

FIGURE 5.9 Temporal evolution of the temperature anomaly (difference between temperature and averaged temperature fields) as measured by HAMSTRAD-Tropo above the altitude of PdM (altitude = 0) in February 2008. Note the gap covering the 48 hours on February 17–18 due to a power failure at the PdM facility.

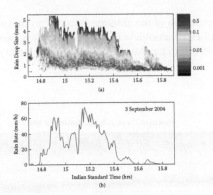

FIGURE 8.11 (a) Contour color map showing the variation of the distribution of the fraction of total number of drops at different sizes during a rain event on September 3, 2004. (b) Variation of rain rate during the same event.

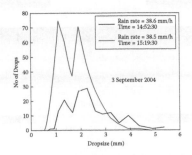

FIGURE 8.12 DSD at two instants of the rain event of September 3, 2005, for identical rain rates indicating a large variation of DSD at different phases of the event.

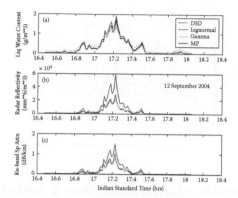

FIGURE 8.13 Variation of different integral rainfall parameters (IRPs) obtained from DSD measurements, a fitted lognormal model, a fitted gamma model, and a Marshall-Palmer (MP) model.

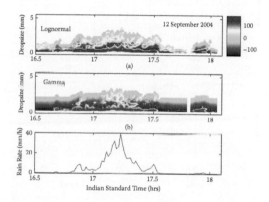

FIGURE 8.14 Contour color map showing the difference between the measured DSD values and the modeled values with (a) lognormal and (b) gamma distributions during the rain event of September 12, 2004.

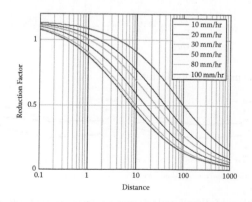

FIGURE 8.17 UK 2003 path reduction factor.

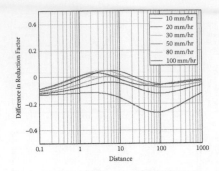

FIGURE 8.18 Modified UK 2003 path reduction factor. It shows a better agreement than proposed in Figure 9.12.

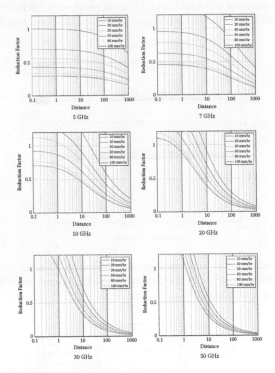

FIGURE 8.19 China path reduction factor.

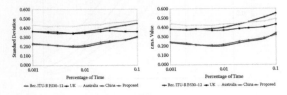

FIGURE 8.20 Testing of Brazil model against terrestrial ITU-R database and other proposed models.

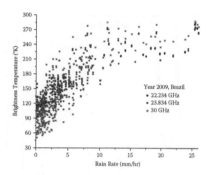

FIGURE 8.23 Variation of brightness temperature with rain intensity.

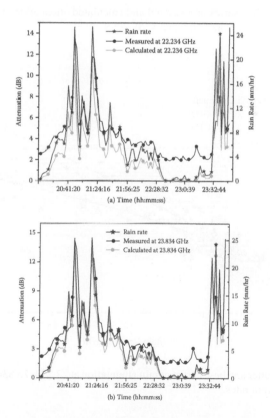

FIGURE 8.24 (a) Time series of measured and calculated attenuation at 22.234 GHz and corresponding rain rates over Brazil on January 20, 2009. (b) Time series of measured and calculated attenuation at 23.834 GHz and corresponding rain rates over Brazil on January 20, 2009.

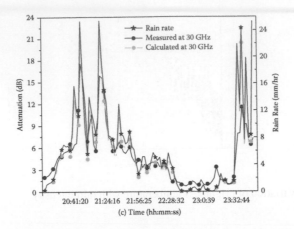

FIGURE 8.24 (c) Time series of measured and calculated attenuation at 30 GHz and corresponding rain rates over Brazil on January 20, 2009.

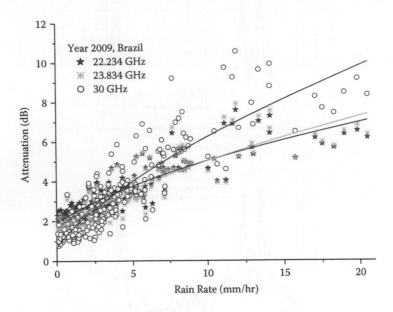

FIGURE 8.25 Scatter and best-fit plot of rain attenuation at 22.234, 23.834, and 30 GHz and the corresponding rain rates.

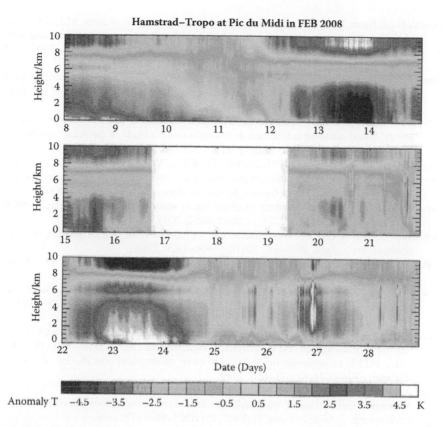

FIGURE 5.9 (See color insert.) Temporal evolution of the temperature anomaly (difference between temperature and averaged temperature fields) as measured by HAMSTRAD-Tropo above the altitude of PdM (altitude = 0) in February 2008. Note the gap covering the 48 hours on February 17–18 due to a power failure at the PdM facility.

REFERENCES

Ackerman, T.P., and G.M. Stokes. The Atmospheric Radiation Measurement Program. *Phys. Today*, 56(1), 38–44, 2003.

Bais, A.F., B.G. Gardiner, H. Slaper, M. Blumthaler, G. Bernhard, R. McKenzie, A.R. Webb, G. Seckmeyer, B. Kjeldstad, T. Koskela, P.J. Kirsch, J. Gröbner, J.B. Kerr, S. Kazadzis, K. Leszczynski, D. Wardle, W. Josefsson, C. Brogniez, D. Gillotay, H. Reinen, P. Weihs, T. Svenoe, P. Eriksen, F. Kuik, and A. Redondas. SUSPEN intercomparison of ultraviolet spectroradiometers. *J. Geophys. Res.*, 106(D12), 12509–12525, 2001.

Brasseur, G.P., J.J. Orlando, and G.S. Tyndall. *Atmospheric chemistry and global change.* 2nd ed. Oxford University Press, New York, 1999.

Cadeddu, M.P., J.C. Liljegren, and A.L. Pazmany. Measurements and retrievals from a new 183-GHz water-vapor radiometer in the Arctic. *IEEE Trans. Geosci. Remote Sensing*, 45(7), 2207–2215, 2007.

Cimini, D., T.J. Hewison, L. Martin, J. Güldner, C. Gaffard, and F.S. Marzano. Temperature and humidity profile retrievals from ground-based microwave radiometers during TUC. *Meteorol. Z.*, 15(5), 45–56, 2006.

Cimini, D., F. Nasir, E.R. Westwater, V.H. Payne, D.D. Turner, E.J. Mlawer, M.L. Exner, and M.P. Cadeddu. Comparison of ground based millimeter-wave observations and simulations in the Arctic winter. *IEEE Trans. Geosci. Remote Sensing*, 47(9), 3098–3106, 2009.

Cimini, D., E.R. Westwater, and A.J. Gasiewski. Temperature and humidity profiling in the Arctic using ground-based millimeter-wave radiometry and 1DVAR. *IEEE Trans. Geosci. Remote Sensing*, 48(3), 1381–1388, 2010.

Cimini, D., E.R. Westwater, A.J. Gasiewski, M. Klein, V.Y. Leuski, and S.G. Dowlatshahi. The ground-based scanning radiometer: A powerful tool for study of the Arctic atmosphere. *IEEE Trans. Geosci. Remote Sensing*, 45(9), 2759–2777, 2007a.

Cimini, D., E.R. Westwater, A.J. Gasiewski, M. Klein, V. Ye Leuski, and J.C. Liljegren. Ground-based millimeter- and submillimeter-wave observations of low vapor and liquid water contents. *IEEE Trans. Geosci. Remote Sensing*, 45(7), 2169–2180, 2007b.

Crewell, S., and U. Lohnert. Accuracy of cloud liquid water path from ground based microwave radiometry. *Radio Sci.*, 38(3), 8042–8052, 2003.

Deblonde, G., and S. English. One-dimensional variational retrievals for SSMIS simulated observations. *J. Appl. Meteorol.*, 42(10), 1406–1420, 2003.

Forkman, P., P. Eriksson, and A. Winnberg. The 22 GHz radio-aeronomy receiver at Onsala Space Observatory. *J. Quant. Spec. Radiat. Transf.*, 77(1), 23–42, 2003.

Hewison, T. 1D-VAR retrievals of temperature and humidity profiles from a ground-based microwave radiometer. *IEEE Trans. Geosci. Remote Sensing*, 45(7), 2163–2168, 2007.

Ide, K., P. Courtier, M. Ghil, and A.C. Lorenc. Unified notation for data assimilation: Operational, sequential, and variational. *J. Meteorol. Soc. Jpn.*, 75(1B), 181–189, 1997.

Liebe, H.J. MPM, an atmospheric millimeter-wave propagation model. *Int. J. Infrared Millimeter Waves*, 10(6), 631–650, 1989.

Mattioli, V., E.R. Westwater, D. Cimini, A.J. Gasiewski, M. Klein, and V.Y. Leuski. Microwave and millimeter-wave radiometric and radiosonde observations in an Arctic environment. *J. Atmos. Ocean. Technol.*, 25(10), 1768–1777, 2008.

Mattioli, V., E.R. Westwater, D. Cimini, J.S. Liljegren, B.M. Lesht, S.I. Gutman, and F.J. Schmidlin. Analysis of radiosonde and ground based remotely sensed PWV data from the 2004 North Slope of Alaska Arctic Winter Radiometric Experiment. *J. Atmos. Ocean. Technol.*, 24(3), 415–431, 2007.

Nedoluha, G., R.M. Bevilacqua, R.M. Gomez, D.L. Thacker, W.B. Waltman, and T.A. Pauls. Ground-based measurements of water vapor in the middle atmosphere. *J. Geophys. Res.*, 100(2), 2927–2939, 1995.

Pommereau, J.-P., and F. Goutail. O_3 and N_2O ground-based measurements by visible spectrometry during Arctic winter and spring 1988. *Geophys. Res. Lett.*, 15(8), 891–894, 1988.

Pyle, J.A., N.R.P. Harris, J.C. Farman, F. Arnold, G. Braathen, R.A. Cox, P. Faucon, R.L. Jones, G. Megie, A. O'Neill, U. Platt, J.-P. Pommereau, U. Schmidt, and F. Stordal. An overview of the EASOE campaign. *Geophys. Res. Lett.*, 21(13), 1191–1194, 1994.

Racette, P.E., E.R. Westwater, Y. Han, A. Gasiewski, M. Klein, D. Cimini, W. Manning, E. Kim, J. Wang, and P. Kiedron. Measurement of low amount of precipitable water vapour using ground based millimeter wave radiometry. *J. Atmos. Ocean Technol.*, 22(4), 317–337, 2005.

Ricaud, P., B. Gabard, S. Derrien, J.-P. Chaboureau, T. Rose, A. Mombauer, and H. Czekala. HAMSTRAD-Tropo, a 183-GHz radiometer dedicated to sound tropospheric water vapor over Concordia Station, Antarctica. *IEEE Trans. Geosci. Remote Sensing*, 48, 3, 2010.

Rodgers, C.D. *Inverse methods for atmospheric sounding: Theory and practice*. World Scientific, Singapore, 2000.

Rosenkranz, P.W. Water vapor microwave continuum absorption: A comparison of measurements and models. *Radio Sci.*, 33(4), 919–928, 1998.

Scherer, M., H. Vömel, S. Fueglistaler, S.J. Oltmans, and J. Staehelin. Trends and variability of midlatitude stratospheric water vapour deduced from the re-evaluated Boulder balloon series and HALOE. *Atmos. Chem. Phys.*, 8(5), 1391–1402, 2008.

Schroeder, J.A., and E.R. Westwater. *User's guide to WPL microwave radiative transfer software*. NOAA Technical Memorandum, ERL WPL-213. National Oceanic and Atmospheric Administration, Boulder, CO, 1991.

Shupe, M.D., T. Uttal, and S. Matrosov. Arctic cloud microphysics retrievals from surface-based remote sensors at SHEBA. *J. Appl. Meteorol.*, 44(10), 1544–1562, 2005.

Smith, E.K. Centimeter and millimeter wave attenuation and brightness temperature due to atmospheric oxygen and water vapour. *Radio Sci.*, 17, 1455–1464, 1982.

Ulaby, F.T., R.K. Moore, and A.K. Fung. *Microwave remote sensing—Active and passive*. Vol. III. Artech House, Norwood, MA, 1986.

Westwater, E.R. Ground-based microwave remote sensing of meteorological variables. In *Atmospheric remote sensing by microwave radiometry*, ed. M. Janssen, 145–213. Wiley, Hoboken, NJ, 1993.

Westwater, E.R., D. Cimini, V. Mattioli, A. Gasiewski, M. Klein, V. Leuski, and J. Liljegren. The 2004 North Slope of Alaska Arctic Winter Radiometric Experiment: Overview and highlights. In *Proceedings of Microw. Spec. Meeting*, San Juan, Puerto Rico, 2006, pp. 77–81.

World Meteorological Organization (WMO). *Scientific assessment of ozone depletion: 2006*. WMO, Geneva, Switzerland, 2006.

6 Radiometric Estimation of Integrated Water Vapor Content

6.1 INTRODUCTION

Out of all meteorological parameters, water vapor and nonprecipitable liquid water, along with ambient temperature, are found to be the most important parameters to control the thermodynamic balance, photochemistry of the atmosphere, sun-weather relationship, and biosphere. Measurements of the vertical and horizontal distribution of water vapor, as well as its temporal variation, are essential for probing into the mysteries of several atmospheric effects. A sizable literature has focused on the complex relationship between water vapor variability and deep convection in the tropics (Sherwood et al., 2009). Unlike higher latitudes, rotational dynamical constraints are weak and precipitation-induced heating perturbations are rapidly communicated over great distances. Water vapor, on the other hand, is highly variable in space and time; its spatial distribution depends on much slower advection processes above the boundary layer and on deep convection itself. Furthermore, deep convection, through vertical transport of water vapor and evaporation of cloud droplets and hydrometeors, serves as the free tropospheric water vapor source. And deep convection is itself sensitive to the free tropospheric humidity distribution through local moistening of the environment, which favors further deep convection, a positive feedback (David et al., 2011). In this context, the ground-based microwave radiometric sensing appears to be one of the suitable solutions for continuous monitoring of ambient integrated atmospheric water vapor. Radiometric data have been extensively used by several investigators (Westwater, 1972; Grody, 1976; Grody et al., 1980; Westwater and Guiraud, 1980; Pandey et al., 1984; Janssen, 1985; Cimini et al., 2007) to determine the water vapor budget.

The first absorption maxima, although weak, in the microwave band occur at 22.234 GHz. So, one can have the choice of exploiting 22.234 GHz on the basis of assumption that the signal-to-noise ratio is largest at this frequency, provided the vertical profiles of pressure and temperature are constant (Resch, 1983). But this does not happen in practice. Pressure and temperature are highly variable parameters of the atmosphere. In this connection, Westwater (1978), showed that the frequency, independent of pressure, lies both ways around the resonance line, i.e., 22.234 GHz. This single-frequency measurement of water vapor was done by Karmakar et al. (1999) at Kolkata (22°N) and at Instituto Nacional de Pesquisas Espaciais (INPE), Cachoeira Paulista (CP) (22°S), Brazil (Karmakar et al., 2010). But incidentally, 22.235 GHz is affected by pressure broadening, although the resonance peak occurs there, and these measurements were not devoid of any influence by the presence of overhead cloud liquid. It has been shown (Simpson et al., 2002) that a zenith-pointing ground-based microwave radiometer measuring sky brightness temperature in the region of 22 GHz is three times more sensitive to the amount of water vapor than the amount of liquid water. However, in the region of 30 GHz, the sky brightness temperature is two times more sensitive to liquid water than that of water vapor, taking into consideration that the sensitivity to ice is negligible at both frequencies, as discussed in Chapter 4.

The measurement of water vapor at Kolkata was done by exploiting a 22.235 and 30 GHz pair (Karmakar et al., 2001). To get the improved desired accuracy of retrieving the vapor budget at the place of choice, two suitable frequency pairs are suggested. In the tropical region, the data on cloud liquid water are quite inadequate, particularly in view of strong seasonal dependence associated with the monsoon (Maitra and Chakraborty, 2009).

Besides this, an attempt has been made in this chapter to show that although the latitudinal occupancy (Kolkata, India, 22°N; CP, Brazil, 22°S) of both the places are the same, the measurement of vapor draws special attention because of entirely different environmental conditions, especially due to the presence of the Amazon Basin in Brazil. The integrated water vapor (IWV) is the vertical column density of atmospheric water vapor. IWV can be determined from microwave radiances measured by multichannel radiometers on the ground or in space. Both the quality and quantity of IWV measurements have rapidly increased during the past 10 to 20 years. Global maps and time series of IWV give evidence for a strong spatiotemporal variability of atmospheric water vapor, playing a key role in weather prediction and climate change research. Hocke et al. (2011) analyze the relationships between microwave radiances and IWV using long-term observations of two radiometers at Bern, Switzerland. The first radiometer (Tropospheric Water Vapour Radiometer (TROWARA)) measures 21 and 31 GHz radiances and permits the accurate retrieval of IWV. The long-term series of the TROWARA have some data gaps that possibly influence the trend analysis. On the other hand, the series of 142 GHz radiance of the second radiometer (Ground-Based Millimeter-Wave Ozone Spectrometer (GROMOS)) is more affected by integrated cloud liquid water (ILW) than the 21 and 31 GHz radiances. The coincident radiometer data of GROMOS and TROWARA are utilized for exploration of the relationship between the 142 GHz radiances of IWV and ILW. The IWV was calculated from the 142 GHz radiance of GROMOS when TROWARA data were not available. Thus, they derived a complete series of IWV above Bern from 1994 to 2009. The combination of both series and the trend analysis were performed by means of multiple linear regressions and bootstrapping. The observations indicate a positive trend up to +10% per decade of IWV in summer and a negative trend of about −15% per decade in winter.

According to Raju et al. (2013) a very unique experimental observation of a ground-based multifrequency microwave radiometer was set to scan the atmosphere in seven elevation angles over coastal Arabian Sea. The spatio-temporal variations of the cloud microphysical parameters during the evolution of a multi-cell convective cumulus system were studied. Humidity and temperature anomalies deduced from the radiometric observation could clearly explain the convective processes like the formation of an intense updraft of moist air, convective heating due to large latent heat energy release, and cooling of the lower atmosphere below 2-km altitude by the down-drafting dry air.

The measurements show the formation of an intense convection in a humid warm atmosphere over a shallow warm ocean (conducive to formation of a waterspout). The life cycles of convective precipitation are of great interest in weather forecasting

Haas et al. (2010) present long-term trends in the amount of atmospheric water vapor at the Swedish west coast. These trends are derived from geodetic very long-baseline interferometry (VLBI), ground-based microwave radiometry, and radiosonde observations. The time span of observations covers 25 years, and the data were collected at the Onsala Space Observatory (VLBI and microwave radiometry) and the Gothenburg-Landvetter Airport (radiosondes). The three techniques detect positive trends in the integrated precipitable water vapor (IPWV) of the order of 0.4–0.6 kg/m² per decade. The IPWV data derived from the three techniques have correlation coefficients of the order of 0.95 and better. However, there is no perfect agreement between the IPWV trends derived by the three techniques. This might partly be explained by different temporal sampling and data gaps. Measurements of down-welling microwave radiation from raining clouds performed (Battaglia, 2011) with the Advanced Microwave Radiometer for Rain Identification (ADMIRARI) radiometer at 10.7, 21, and 36.5 GHz during the Global Precipitation Measurement Ground Validation "Cloud Processes of the Main Precipitation Systems in Brazil: A Contribution to Cloud Resolving Modeling and to the Global Precipitation Measurement" (CHUVA) campaign held in Brazil during 2010–2011 at different latitudes represent a unique type equation test bed for understanding three-dimensional (3D) effects in microwave radiative transfer processes. While the necessity of accounting for geometric effects is trivial given the slant observation geometry (ADMIRARI was pointing at a fixed 30° elevation angle), the polarization signal (i.e., the difference between the vertical and horizontal brightness temperatures) shows the ubiquitousness of positive values at both 21.0 and 36.5 GHz in coincidence with high brightness temperatures. This is a genuine and unique microwave signature of radiation side leakage that cannot be explained in a 1D radiative transfer frame, but necessitates the inclusion of three-dimensional scattering effects.

6.2 SINGLE-FREQUENCY ALGORITHM FOR WATER VAPOR ESTIMATION

The vertical and horizontal distributions of water vapor will vary in both space and time. Resch (1983) calculated the atmospheric emission around 22 GHz for two different vertical distributions of water vapor. The pressure distribution in a standard atmosphere used there is a surface temperature of 30°C and standard lapse rate. For a constant relative humidity of 81.6% and for the height region 0–1 km (Figure 6.1(a)), and for 99% humidity and for a height region of 1.0–3.8 km (Figure 6.1(b)), the brightness temperatures were calculated. The high-altitude vapor clearly shows a sharper line than the low-altitude profile, although the delays in both cases were the sane. Hence, it may be suggested that if one wishes to maximize signal from a given amount of water vapor, then clearly the observation should be carried at 22.235 GHz. It is also clear from the figure that a single-frequency measurement of brightness temperature near the half power point of the line profile would provide the most accurate estimates. But, it is to be remembered that the algorithm developed below should be valid only in the case of a cloudless and clear sky.

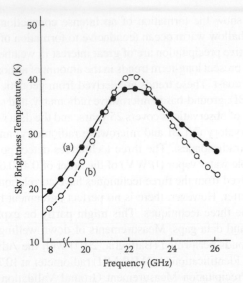

FIGURE 6.1 Line profiles of atmospheric water vapor for two different vertical distributions in standard atmosphere. (a) $RH = 81.6\%$ for $0 < H < 1000$ m (open circles). (b) RH = 99% for $1000 < H < 3800$ m (filled circles).

The absorption (cm^{-1}) of a water vapor molecule at 22.235 GHz is given by Bhattacharya (1985) as

$$\Gamma = 3.24 * 10^{-4} \frac{P exp\left[-\dfrac{644}{T}\right]\gamma^2}{T^{3.125}} \times \left[1 + 0.0147 \frac{\rho T}{P}\right]$$

$$\times \left[\frac{1}{(\gamma - 22.235)^2 + (\Delta\gamma)^2} + \frac{1}{(\gamma + 22.235)^2 + (\Delta\gamma)^2}\right] + 2.55 \times 10^{-8} \frac{\rho\gamma^2\Delta\gamma}{T^{\frac{3}{2}}}$$

(6.1)

where γ is the frequency in GHz, T is the kinetic temperature in K, and $\Delta\gamma$ is the pressure-broadened line half-width parameter and is given by

$$\Delta\gamma = 2.58 \times 10^{-3} \left[1 + 0.0147 \frac{\rho T}{P}\right] \frac{P}{\left(\dfrac{T}{318}\right)^{0.625}} GHz$$

(6.2)

where P is the total pressure ($mm\ of\ Hg$) and $\rho(gm^{-3})$ is the water vapor density.

After simplification, Equations 6.1 and 6.2 yield

$$\Gamma = 17.92 \frac{\rho exp\left[\dfrac{-644}{T}\right]}{PT^{1.875}} \times \left[1 + 0.0147 \frac{\rho T}{P}\right]^{-1}$$

$$+ 11.91 \times 10^{-7} \left[1 + 0.0147 \frac{\rho T}{P}\right] \frac{\rho}{T^{2.125}} cm^{-1}$$

(6.3)

In Equation 6.3, the first term is the resonance term and the second term is the nonresonance term. Now, considering the typical surface parameters, $P = 750$ mm of Hg, $T_0 = 300$ K, and $\rho_0 = 25$ g^{-3}, we find the contribution of the nonresonance part is of the order of 1% of the resonant part. Hence, the nonresonance part is neglected in comparison to the resonant part. Thus, we are left with

$$\lambda = 17.92 \frac{\rho\exp\left[\dfrac{-644}{T}\right]}{PT^{1.875}} \times \left[1 + 0.0147\frac{\rho T}{P}\right]^{-1} \text{cm}^{-1} \tag{6.4}$$

Moreover, using the same data, we find that the value of 0.0147 $\rho T/P = 0.145$ is much less in comparison to the first term, and hence is neglected. Now, converting Equation 6.4 into dB km^{-1}, we get

$$\lambda = 17.92[\log_{10}\rho \times 10^6] \times \frac{\rho\exp\left[\dfrac{-644}{T}\right]}{PT^{1.875}}$$

$$= 7.78 \times 10^6 \times \frac{\rho\exp\left[\dfrac{-644}{T}\right]}{PT^{1.875}}$$

$$= 7.78 \times 10^6 \times \rho(T) \times F(T) \times \frac{T^{0.52699}}{P} \, dBkm^{-1} \tag{6.5}$$

where $F(T) = \dfrac{\exp\left[-\dfrac{644}{T}\right]}{2.40199}$ is an implicit function of temperature (Karmakar, 1989).

The range of temperatures (T) over a tropical location like Kolkata is such that $F(T)$ represents a slowly varying function, as depicted in Figure 6.2. It is found there that the monthly variation of T makes $F(T)$ a slowly varying function over the range of temperatures, which is considered to be a typical variation of the parameter over Kolkata during the full calendar year. Now, referring to Figure 6.2, we have taken the liberty to take the value of $F(T)$ equal to 1.3×10^{-7} (average value). Hence, Equation 6.5 reduces to

$$\lambda = 7.78 \times 10^6 \times 1.3 \times 10^{-7} \times \frac{\rho T^{0.52699}}{P} \, dBkm^{-1}$$

$$= 1.0114 \times \frac{\rho T^{0.52699}}{P} \, dBkm^{-1}$$

On integration, we get

$$A = 1.0114 \int_0^\alpha \frac{\rho T^{0.52699}}{P} \, dh \, (dB) \tag{6.6}$$

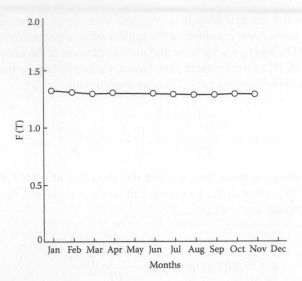

FIGURE 6.2 Monthly variation of $F(T)$. It is an implicit function of atmospheric temperature, T, in Kolkata.

Here A represents the total atmospheric water vapor attenuation in dB, along a vertical path.

To carry out the integration, according to Hess (1959) and Braunt (1947), the atmosphere is assumed to be of a constant lapse rate. In an atmosphere having a constant lapse rate, the relation between temperature (T) and pressure (P) is given by Poission's equation, which may approximately be written as

$$T = T_0 \left[\frac{P}{P_0} \right]^{R\beta/g} \tag{6.7}$$

$$P = P_0 \exp\left(\frac{-h}{H_\rho} \right) \tag{6.8}$$

$$\rho = \rho_0 \exp\left(-\frac{h}{H_\rho} \right) \tag{6.9}$$

where R is the gas constant, β is the lapse rate, T_0, P_0, and ρ_0 are the surface parameters, H_P and H_ρ are the pressure scale height and water vapor scale height, respectively, and g is the acceleration due to gravity. Now, from Equation 6.6,

$$A = 1.0114 \times \frac{\rho_0 T_0^{0.52699}}{P_0} \int_0^\alpha \exp(-H_1 h) dh$$

$$= 1.0114 \times \frac{\rho_0 T_0^{0.52699}}{P_0 H_1}$$

where

$$H_1 = \frac{1}{H_\rho} - \frac{1-(0.52699R\beta/g)}{H_\rho} \tag{6.10}$$

For Kolkata, $\beta = 0.7509$ K/100 m, the average adiabatic lapse rate for 1.5 to 7 km (data taken from Civil Aviation Department, Kolkata Airport; $R = 2.9$ units and $H_P = 8$ km.

From Equations 6.8 and 6.9 and substituting R, H_P, and β we get

$$\frac{P_0 A}{H_\rho} = 1.0114 \times T_0^{0.52699} \times \rho_0 + P_0 \times 0.1103\lambda$$

$$H_\rho = \frac{P_0 A}{1.0114 \times T_0^{0.52699} \times \rho_0 + P_0 \times 0.1103\lambda} \text{ km} \tag{6.11}$$

where A is the zenith attenuation (dB) to be measured by using the 22.235 GHz radiometric data.

However, according to Allnutt (1976), the zenith attenuations may be calculated from the radiometric output in the form of brightness temperature by using the equation

$$A = 10 log_{10} \frac{T_m - T_{cos}}{T_m - T_a(f)} \tag{6.12}$$

where $T_a(f)$ represents the brightness temperature at a frequency of choice. Here T_m is the mean atmospheric temperature, which is eventually found to be dependent on ground temperature, obeying the relationship $T_m = C(f) T_s$. Here C is considered to be a frequency-dependent term. Karmakar (1989) found $C(22.235) = 0.95$. So with the available data for T_s (surface temperature), the values of T_m may be calculated for different occasions. Now with the accepted values of $T_{cos} = 2.75$ K (noise temperature due to cosmic background), the attenuation values from the measured radiometric brightness temperature T_B (refer to Equation 6.12) may be calculated. Hence by using Equation 6.11, the appropriate values of water vapor scale height may be determined.

6.2.1 ATTENUATION AT 22.234 GHz

The variation of calculated attenuation (dB) and the corresponding monthly variation of surface water vapor density (g m^{-3}) over Kolkata (22°N), India, are shown in Figure 6.3. A time series of attenuation in CP (22°S), Brazil, is shown in Figure 6.4. The monthly variation of calculated and measured brightness temperature over Kolkata is shown in Figure 6.5. Both figures show that the brightness temperatures and water vapor density bear a maximum during the months of July and August over Kolkata. This is presumably due to maximum abundances of water vapor in these months. Although the latitudinal occupancy of both places is the same, the measurement of water vapor over Brazil draws special attention because of entirely different environmental conditions, especially due to the Amazon Basin. Looking at

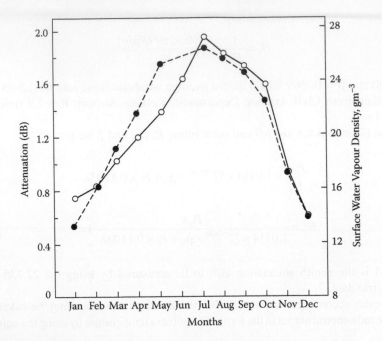

FIGURE 6.3 Monthly variation of calculated attenuation (dB) and corresponding monthly variation of surface water vapor density (g^{-3}) over Kolkata (22°N).

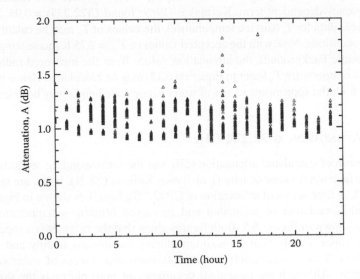

FIGURE 6.4 Time series of 22.234 GHz attenuation (dB) in clear sky (without clouds) in CP, Brazil (22°S).

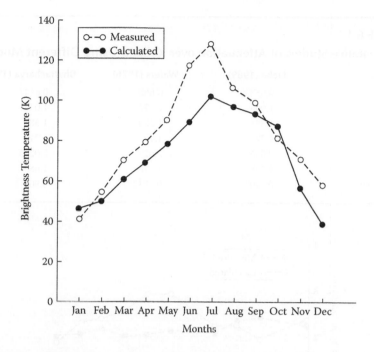

FIGURE 6.5 Monthly variation of calculated and measured brightness temperature for clear sky in Kolkata.

Figure 6.4, it should be mentioned that the radiometer has the provision to measure the height of the cloud base, which is identified as 0 for no cloud and 1 for cloudy conditions in the zenith direction. For cloudless conditions, Figure 6.4 shows that the attenuation always remains within 1.0–1.5 dB, except at around 14 through 18 hours, when the attenuation reaches a value of more than 1.5 dB. This might be due to the fact that in the afternoon the water vapor concentration goes to a maximum. Usually after this span of time rain starts, provided the saturation occurs.

A comprehensive study has been made to compare the calculated values of attenuations obtained by different authors (Liebe, 1989; Waters, 1976; Bhattacharya, 1985). The results are summarized in Table 6.1. Now if we look at Figures 6.3 and 6.4, we will see some discrepancies between measured and calculated values. This might need the inclusion of an empirical, nonresonant correction that depends on the square of water vapor density. But still, the origin of this effect is not properly understood. The possible reason has been sought in terms of hydrogen bonding of the water molecule to form a dimer (Bohlander et al., 1980) or of the water molecules together or in terms of errors in the line shape used in the calculation (Gibbins, 1986).

6.2.2 WATER VAPOR SCALE HEIGHT BY DEPLOYING 22.234 GHz RADIOMETER

Karmakar et al. (1999, 2011) reported the variational pattern of water vapor scale height (refer to Equation 6.11) in Kolkata, India, and CP, Brazil (see Figures 6.6 and 6.7). It is also clear from Figure 6.7 that the water vapor scale height values

TABLE 6.1

Comparative Studies of Attenuation over Kolkata Using Different Models

Month	Liebe (1989)	Waters (1976)	Bhattacharya (1985)
January	0.65339	0.665	0.6143
March	1.065	1.072	1.035
May	1.391	1.390	1.386
June	1.724	1.703	1.716
July	2.036	2.027	2.0235
September	1.7084	1.676	1.683
November	0.9403	0.9533	0.905

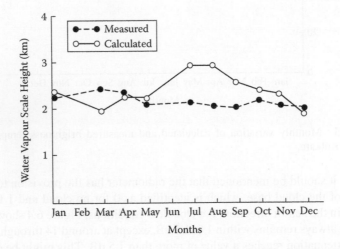

FIGURE 6.6 Monthly variation of calculated and measured water vapor scale height over Kolkata (22°N).

attain a minimum around 17 hours of the day with no overhead cloud. However, the presence of cloud may produce a deviation in this pattern. It may also happen that the suspended nonprecipitable liquid water molecules in the atmosphere, those which are to be evaporated by taking the appropriate amount of latent heat from the atmosphere, produce a little deviation in determination of water vapor scale height. Moreover, from Figure 6.7 it is observed that the water vapor scale height, on average, during the whole day is 2 km. Sen et al. (1989) found that for a time resolution between 12 and 24 hours, the correlation between the variation pattern of integrated water vapor content W (kgm^{-2}) and water vapor density at 2 km height is very good. Hence, it is concluded that any transportation of water vapor from the surface to the altitude of 2 km must have a negligible effect within this timescale. If the timescale is increased to 48 hours, the integrated water vapor content is poorly correlated with that around 2 km height. The difference in behavior in this type of variation for a

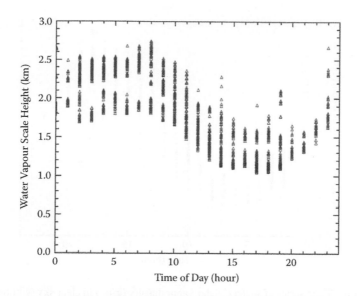

FIGURE 6.7 A time series of water vapor scale height on April 15 over Brazil (22°S) during clear sky conditions.

short (12–24 hours) and a long (48 hours) scale suggests that the transportation of water vapor to high altitudes occurred within a timescale greater than 24 hours.

6.2.2.1 Water Vapor Density and Vapor Pressure

The time variations of surface water vapor density and surface water vapor pressure are shown respectively in Figures 6.8 and 6.9. Radiosonde analyses in deriving water vapor density at CP, INPE, Brazil, reveal that from 15 through 18 hours, during no cloud condition, it attains a maximum value of about 38 g/m³, and subsequently water vapor pressure also shows a similar trend of achieving of about 55 *hPa* pressure at around 17 hours. It is also interesting to note that in the early morning (Brazil local time) hours the surface water density takes the minimum value of about 15 g/m³, and subsequently water vapor pressure becomes 20 *hPa*.

6.2.3 Integrated Vapor Content by Deploying 22.234 GHz Radiometer

Now remembering Equation 6.11, we write water vapor content:

$$W = 10^3 \times H_\rho \times \rho_0 \tag{6.13}$$

An attempt has been made to present derived water vapor content from the radiometric measurement variation with the directly obtained brightness temperature from radiometer at 22.234 GHz during clear weather. This is presented in Figure 6.10. Regression analysis of the scatter plot shows the best linear equation over INPE, Brazil (22°S), is

$$W = 478.45 T_B + 9574.36 \tag{6.14}$$

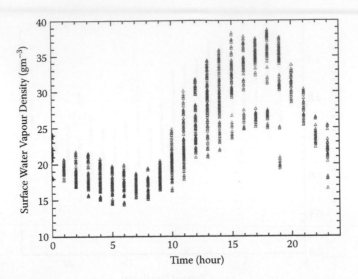

FIGURE 6.8 Time series of surface water vapor density (g/m³) in clear sky (without clouds) in CP, Brazil (22°S).

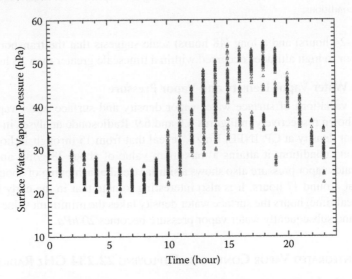

FIGURE 6.9 Time series of surface water vapor pressure (hPa) in clear sky (without clouds) in CP, Brazil (22°S).

However, from the estimated regression equation over Kolkata (22°N, it was found that the water vapor content is related to brightness temperature as (Karmakar, 1989)

$$W = 612.0 T_B + 16400 \qquad (6.15)$$

and that over Delhi (28.38°N) as (Bhattacharya, 1985)

$$W = 588.23 T_B - 2110 \qquad (6.16)$$

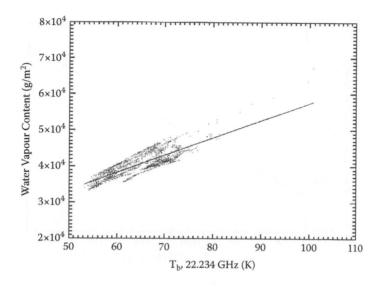

FIGURE 6.10 Regression analysis of water vapor content (g/m²) and brightness temperature (K) in CP, Brazil.

To provide more clarity for Figure 6.10, it was redrawn at a larger scale of brightness temperature and presented in Figure 6.11. Now to get a comparative study between the calculated (using the radiosonde data available from balloon flight at a nearby station, which is 100 m away from the site of the radiometer) values of water vapor content, y, and the measured (using the radiometer) water vapor content, x, it is presented pictorially (Figure 6.12). This shows the relationship that exists between the two is

$$y = 0.4218x + 31.31 \qquad (6.17)$$

The r.m.s. error between these two types of studies is 0.49 kg/m². This difference might be due to spatial and temporal baseline and sensor accuracy, especially while calculating the absolute humidity profile using radiosonde data.

In the present case, the integrated water vapor content has been measured by deploying a 22.234 GHz radiometer. We have assumed that there is no such appreciable variation of pressure and temperature, but that is not the case. In principle, when pressure and temperature change, as occurs with height in the troposphere, operation at the line center is not optimal (Hogg et al., 1983). The frequency is to be selected a little away from the resonance line, which could primarily sense the vapor. In the next section an effort is given to use multifrequency radiometric data to improve the accuracy. These frequencies should be independent of particular distributions of water vapor with height. The strong and sufficient/significant similarity between the calculated and measured values of water vapor content suggests that the radiometer can be used for continuous monitoring of water vapor.

However, in order to reduce the level of existing uncertainties in climate forecast considering the primary role of water vapor, improved modeling of the atmosphere

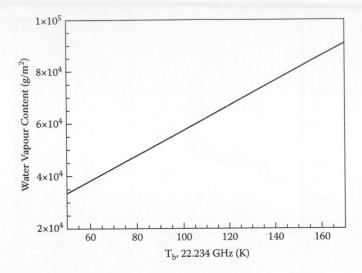

FIGURE 6.11   Same as Figure 6.10, but redrawn in larger scale.

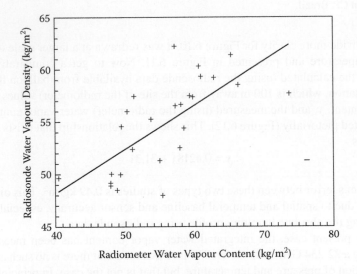

FIGURE 6.12   Scatter plot of measured and calculated water vapor content (kg/m²) in Brazil.

is urgently required on a global basis (Cartalis and Varotsos, 1994; Varotsos et al., 2001; Kondratyev and Varotsos, 2001a, 2001b).

In this regard special attention has been paid to the assessment of greenhouse warming versus aerosol cooling. However, among the most important aspects of the atmospheric pollution problem are the anthropogenic impacts on the stratospheric ozone layer, the related trends of the total ozone content drop, and the solar ultraviolet radiation enhancement at the earth's surface level, which cause various dangers for man and ecosystems (Varotsos, 2004).

6.3 DUAL-FREQUENCY ALGORITHM FOR WATER VAPOR ESTIMATION

It has already been discussed that the selection of frequency at 22.234 GHz for the measurement of water vapor in the presence of cloud is not optimal because the nonprecipitating cloud liquid contaminates the measured vapor. Besides this, the choice of 22.234 GHz is pressure dependent and does not reflect the true values. This has been discussed in Chapter 4 (see also Figure 4.1). There it was concluded that 23.834 GHz is such a pressure-independent frequency for atmospheric water vapor measurement. Keeping these in view, it was interesting to use dual-frequency radiometric measurement of water vapor at Instituto Nacional de Pesquisas Espaciais (INPE), Cachoeira Paulista (CP), Brazil (22°S), exploiting the 23.834 and 30 GHz pair. The measurement of water vapor at Kolkata was also done by exploiting the 22.235 and 30 GHz pair (Karmakar et al., 2001). Taking advantage of the presence of more channels in the radiometer in the water vapor band, as described in Chapter 2, two more frequency pairs were used to check the desired accuracy of retrieving integrated vapor at the place of choice. Moreover, an attempt has also been made to use a 23.034 pair, a few GHz away from the peak line frequency (refer to Figure 4.1), and 30 GHz pair.

In the tropical region, the data on cloud liquid water are quite inadequate, particularly in view of strong seasonal dependence associated with the monsoon (Maitra and Chakrabarty, 2009). In the present context, liquid water (LWC) has also been obtained from dual-frequency radiometric measurements at a tropical location to indicate its variational pattern. Besides this, it was mentioned earlier that although the latitudinal occupancies (Kolkata, India, 22°N; CP, Brazil, 22°S) of both places are the same, the measurement of vapor draws special attention because of entirely different environmental conditions, especially due to the presence of the Amazon Basin in Brazil.

6.3.1 THEORETICAL BACKGROUND

Atmospheric attenuation in the microwave region is mainly due to water vapor, liquid water, and gaseous particles (mainly oxygen). Among these, water vapor plays a major role in attenuating the microwave signal during its propagation through the ambient atmosphere. A microwave signal also gets attenuated in the presence of clouds containing liquid water. Gaseous oxygen, though not as intense as water, contributes to the attenuation. Now, the total attenuation in dB is written mathematically as

$$A_T = A_V(f) + A_L(f) + A_O(f) \tag{6.18}$$

Here A_T is the total attenuation, A_V is the attenuation due to water vapor, A_L is the attenuation due to nonprecipitable cloud liquid water, and A_O is the attenuation due to oxygen. The total attenuation for the higher frequency (f_2) can be formulated as

$$A_T(f_2) = A_V(f_2) + A_L(f_2) + A_O(f_2) \tag{6.19}$$

Similarly, the total attenuation involving frequency f_1 (the lesser one) can be written as

$$A_T(f_1) = A_V(f_1) + A_L(f_1) + A_O(f_1) \tag{6.20}$$

Attenuations due to cloud liquid at the two frequencies are related as (Karmakar et al., 2001)

$$A_L(f_2) = K_1 A_L(f_1) \tag{6.21}$$

where

$$K_1 = \left(\frac{f_2}{f_1}\right)^2$$

Multiplying Equation 6.20 by K_1 we get

$$K_1 A_T(f_1) = K_1 A_L(f_1) + K_1 A_O(f_1) + K_1 A_V(f_1) \tag{6.22}$$

From Equations 6.19 and 6.22 and rearranging we get

$$[K_1 A_T(f_1) - A_T(f_2)] - [K_1 A_O(f_1) - A_O(f_2)] = K_1 A_V(f_1) - A_V(f_2) \tag{6.23}$$

Now we choose the left-hand side of Equation 6.23, to be written as

$$Z_v = [K_1 A_T(f_1) - A_T(f_2)] - [K_1 A_O(f_1) - A_O(f_2)] \tag{6.24}$$

Here we see that the term Z_v contains only the vapor part and dry part and can be termed the attenuation part free from liquid attenuation.

A statistical regression analysis between the calculated values of water vapor attenuation $A_V(f_1)$ and $A_V(f_2)$ by using radiosonde data reveals that they are related as

$$A_V(f_2) = K_2 A_v(f_1) \tag{6.25}$$

where k_2 is the regression constant

Again, involving Equation 6.21, and progressing similarly to the case of water vapor, we get

$$Z_L = [K_2 A_T(f_1) - A_T(f_2)] - [K_2 A_O(f_1) - A_O(f_2)] \tag{6.26}$$

6.3.2   RADIOSONDE DATA ANALYSIS OF VAPOR ESTIMATION

We define the mass of water vapor content in a cylindrical column of infinite length and of 1 m² cross-sectional base area as integrated precipitable water vapor (IPVP), expressed in kg/m². Similarly, liquid water is the depth of water that could be collected from a column of cloud liquid water droplets, which can also be expressed in kg/m².

This measurement cannot be done analytically. So it is accomplished by constructing a database of water vapor content, liquid water content, and brightness temperature values. The amount of water vapor contained in the atmosphere is a function of several meteorological parameters, but it especially depends on the atmospheric temperature. The primary parameter of interest in finding the water vapor density is the partial water vapor pressure (hPa), which is given by the relation (Moran and Rosen, 1981)

$$e = 6.10 \exp\left\{25.228\left(1-\left(\frac{273}{T_D}\right)\right)-\left(5.31 \log\left(\frac{T_D}{273}\right)\right)\right\} \tag{6.27}$$

where T_D is the dew point in K.

But, during saturation this water vapor pressure can be formulated as

$$e_s = 6.112 \exp\frac{(17.502 \times t)}{t+240.97}\,\text{hPa} \tag{6.28}$$

where t is the atmospheric temperature in °C.

Now, one can take the liberty to consider that water vapor approximately behaves as an ideal gas, where each mole of gas obeys an equation of state that can be written as

$$\rho_v = \frac{e}{R_w}T \tag{6.29}$$

where ρ_v is the water vapor density (kg/m³) and e is the partial pressure of water vapor (hPa). R_w is the gas constant for water vapor, i.e., $R_w = R/m_w$, where R is the universal gas constant, m_w is the mass of 1 mole of vapor, and T is the absolute temperature (K). Here $R = 8.135$ J/mole-K and $m_w = 18$ g. Substituting all these above values into Equation 6.29, we find the water vapor density (g/m³).

Each radiosonde ascent provides a profile of atmospheric pressure, altitude, temperature, and dew point temperature data. In the standard data format used, all four quantities are recorded generally, at 15 specified values of atmospheric pressure. Intermediate values may be recorded whenever there are significant changes of pressure or temperature at corresponding altitudes. Those can be determined by interpolation, assuming an exponential profile between the relevant pressure values. Vertical profiles that match the slab heights required by Liebe's model are then obtained by further interpolation. It is to be noted that the atmosphere may be divided into 0.2 km thick slabs, and above 5 km it has been assumed as 1 km thick. The pressure, temperature, and dew point temperature at each height are used to derive relative humidity, air density, and vapor pressure for each slab. From these we may obtain the water vapor density profile, and hence vapor content, by integration with respect to height.

In the present context, we are interested in finding the integrated water vapor content, defined earlier, which can be written as

$$\text{V(kg/m}^2) = \int_0^\infty \rho_v(h)\,dh \tag{6.30}$$

Water vapor absorption coefficients (α_v, dB/km) at the corresponding heights are obtained using Leibe's MPM model (Liebe, 1985). The data inputs are temperature, pressure, and R_H, which is given by

$$R_H = \frac{e}{e_s} \times 100$$

6.3.3 RADIOSONDE DATA ANALYSIS OF CLOUD ATTENUATION

We are usually accustomed to different types of cloud discussed by several authors from time to time, along with their generation and formation mechanism. Among those types of clouds, we were concerned with clouds when the radiometric temperatures were a little higher in the presence of the cloud. These types of cloud sometimes extend well above the cirrus type and penetrate a few thousand feet into the stratosphere. The water content of clouds normally increases upward to a maximum in the vicinity of the melting level. These types contain exceedingly high vertical velocities, and in that case the presence of hail above the tropopause is also possible.

But, it is to be pointed out that such steady-state conditions with respect to water distribution in the cloud do not exist when updrafts equal or exceed the fall velocity of particle as rain. Such updrafts exist only locally for periods of about 5–15 minutes in convective activity (Adams et al., 2011), and hence can lead to a high local concentration of water. However, the calculation of the columnar liquid water content of clouds from radiosonde measurements is based on the model proposed by Salonen et al. (1991).

The cloud detection is performed by using the critical humidity function, defined as follows:

$$U_c = 1 - \sigma\alpha(1-\sigma)\left[1 + \beta(\sigma - 0.5)\right] \tag{6.31}$$

Here $\sigma = \dfrac{p}{p_0}$ = ratio between the atmospheric pressure at the considered level and the pressure at the ground, where $\alpha = 1.0$ and, $\beta = \sqrt{3}$ (an empirical constant), as proposed by Salonen (1991). In Silverman and Sprague's model used earlier by Karmakar et al. (2001) at Kolkata, it was presumed that the distribution of liquid water within the cloud thickness ranging from 660 m to 3.4 km was not uniform (no actual profile was mentioned) without having any prior knowledge of cloud thickness over Kolkata. But in the Salonen model, the cloud thickness can easily be evaluated by using Figure 6.13, wherein the cloud base and top were found to be 1.5 and 3.0 km and distributions were taken as uniform. This helps us to provide the proper integration limit while retrieving vapor separating liquid content.

Besides these, in the Salonen model no hail or ice cloud was taken into consideration that favors the choice of this model over Brazil. Salonen proposed this model for cloud water density in terms of temperature and height from cloud base that can be used to obtain the liquid water density profile within clouds.

The existence of clouds at the certain level was taken into consideration when the relative humidity at a particular height was greater than the U_C, as mentioned in Equation 6.31. Again, it is presumed that within the cloud layer the water density ρ_C of

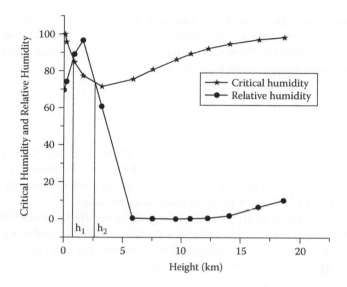

FIGURE 6.13 Curve illustrating the method to find the heights between which clouds exit.

any slice of the upper air sounding is a function of the air temperature, t (°C), and of the layer height, h (m). The existence of cloud at a certain level was taken into consideration when the relative humidity at a particular height was greater than the U_c. The plot of the vertical profile of $U_c(h)$ and R_H for a particular day, April 10, 2009, over Brazil, for the sake of clarity, is presented in Figure 6.13. The cloud base height and the cloud top height are found by the interpolation method, as shown in the same figure.

Here h_1 is the base height of the cloud where the RH curve crosses the $U_c(h)$ curve for the first time. The cloud top h_2 is the point on the curve where the curve $U_c(h)$ again crosses the R_H curve. Again, it is presumed that within the cloud layer the water density ρ_c of any slice of the upper air sounding is a function of the air temperature, $t°C$, and of the layer height, h (m). It is given by

$$\rho_C (t,h) = \rho_0 \exp(Ct) \left(\frac{h - h_b}{h_r} \right) \rho_w \tag{6.32}$$

where $\rho_0 = 0.17$ (g/m³), $C = 0.04$ (/°C) and is a temperature-dependent factor, $h_r = 1500$ (m), and h_b = cloud base height (m).

And $\rho_w (t)$ is the liquid water fraction, approximated by

$$\rho_w (t) = 1 \quad \text{if } 0°C < t$$

$$\rho_w (t) = 1 + t/20 \quad \text{if } -20°C < t < 0°C \tag{6.33}$$

$$\rho_w (t) = 0 \quad \text{if } t < -20°C$$

Here it should be mentioned that Equation 6.33 was the key equation derived by Salonen et al. (1991). This model was also used by Maitra and Chakrabarty (2009) for Kolkata by using the radiosonde data only.

The calculation of both cloud base and top heights may be worked out from the ground by linear interpolation. The columnar liquid water content, L, can be found by adding the contributions from all the layers within the clouds that contain water. The total liquid water content is given by

$$L\left(\frac{\text{kg}}{\text{m}^2}\right) = \int_{h1}^{h2} \rho_C(h)\,dh \tag{6.34}$$

Here it is to be noted that over Brazil (22°S), the h_2 values were different for different days. But the maximum value of h_2 was found as 3 km. So it is conclusively decided that over Brazil, the cloud never was extended up to the stratosphere, and the question of formation of hail above the tropopause does not arise.

For clouds consisting entirely of small droplets, generally less than 0.01 cm, the Rayleigh approximation is valid for frequencies below 200 GHz, and it is possible to express the attenuation in terms of the total water content per unit volume. Thus, the specific attenuation within a cloud can be written as

$$\alpha_c = K_L \rho_c \tag{6.35}$$

where α_c = specific attenuation (dB/km) within the cloud, K_L = specific attenuation coefficient (dB/km)/(g/m³), and ρ_c = liquid water density in the cloud (g/m³).

For a mathematical model based on Rayleigh absorption for nonprecipitating clouds, we assume the liquid absorption depends only on the total liquid amount and not on the drop size distribution. The Rayleigh approximation seems to be valid when the scattering parameter is much less than 1, which uses a double-Debye model for the dielectric permittivity, $E(f)$, of water. This idea can be used to calculate the value of K_L for frequencies up to 1000 GHz where

$$K_L = \frac{0.819 f}{\varepsilon''(1+\eta^2)} \tag{6.36}$$

and f is the frequency (GHz) and $\eta = \dfrac{2+\varepsilon'}{\varepsilon''}$

The complex dielectric permittivity of water is given by

$$\varepsilon''(f) = \frac{f(\varepsilon_0 - \varepsilon_1)}{f_p\left[1+\left(\dfrac{f}{f_p}\right)^2\right]} + \frac{f(\varepsilon_1 - \varepsilon_2)}{f_s\left[1+\left(\dfrac{f}{f_s}\right)^2\right]}$$

and

$$\varepsilon'(f) = \frac{\varepsilon_0 - \varepsilon_1}{\left[1+\left(\dfrac{f}{f_p}\right)^2\right]} + \frac{(\varepsilon_1 - \varepsilon_2)}{\left[1+\left(\dfrac{f}{f_s}\right)^2\right]} + \varepsilon_2$$

Here $\varepsilon_0 = 77.6 + 103.3\,(\theta - 1)$, $\varepsilon_1 = 5.48$, $\varepsilon_2 = 3.51$, $\theta = 300/T$, and T is the ambient temperature (K). The principal and secondary relaxation frequencies are $f_p = 20.09 - 142\,(\theta - 1) + 294\,(\theta - 1)^2$ GHz and $f_s = 590 - 1500\,(\theta - 1)$ GHz. This specific attenuation when integrated over the vertical path gives the total attenuation in dB. The equation involved is

$$A_c\,(dB) = \int_0^\alpha \infty(h)\,dh \tag{6.37}$$

Brightness temperatures at different times of different months are made available from the radiometer data where from the total attenuation may be obtained using the relation given by Allnutt (1976) (refer to Equation 6.12). Now the measured total attenuation values (A_T) are used to obtain the values of Z_L using Equation 6.26. Liquid water content may be calculated using Equation 6.39. The values of p_L and q_L may be obtained from the radiometric record for the desired period.

Referring back to Equation 6.24, we see that the term Z_V contains only the vapor part and dry part and can be termed the attenuation part free from liquid attenuation.

A statistical least square fitting may be adopted between Z_V and calculated values of V obtained by using radiosonde data and the measured values obtained from radiometer data for the f_1 and f_2 frequency pair. Now the measured radiometric brightness temperature for a pair of frequencies and Equation 6.24 yield

$$V\left(\frac{kg}{m^2}\right) = Z_v p_v + q_v \tag{6.38}$$

Here p_V and q_V are the regression coefficients.

Similarly, using Equations 6.9 and 6.15, we find

$$L\left(\frac{kg}{m^2}\right) = Z_L p_L + q_L \tag{6.39}$$

A time series of attenuation values at 23.834 and 30 GHz has been presented in Figure 6.14, over Brazil. It shows that as time passes toward the end of April the attenuation values decrease, and the attenuation at 23.834 is three times larger than that obtained at 30 GHz. The measured values of attenuations for the frequency pair are then substituted in Equation 6.24 to find Z_V. Here it may be noted that the second term in the parentheses of Equation 6.25 is the dry term (A_0), which is mainly due to oxygen present in the ambient atmosphere. This part is calculated by using the MPM model (Liebe, 1985), where the input parameters are only ambient temperature, pressure, and humidity of the atmosphere. It was also found that the variation of A_0 is almost negligible over the place in question. But to achieve good accuracy, this part has been included in the present context to get the values of the regression constants, like p_V and q_V (see Table 6.2). The same idea as noted above has been taken into consideration while finding Z_L (see Table 6.2).

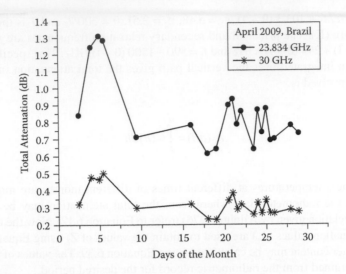

FIGURE 6.14   Time series of attenuation at 23.834 and 30 GHz.

TABLE 6.2
Different Regression Constants Used in the Algorithm

Regression Constants	23.834 and 30 GHz Frequency Pair
K_1	1.5843
p_v	31.0428
q_v	8.06235
K_2	0.43731
p_L	4.0701
q_L	-2.8409×10^{-4}

It should be mentioned that over Brazil the monsoon season prevails from January through April. Now, the regression coefficients for the month of April along with the values of Z_V by exploiting the dual-frequency pair have been used for the measured values of integrated water content (kg/m²). A time series of the measured values of vapor content is presented and compared with those obtained by using the radiosonde data (Figure 6.15). This shows the vapor content is at a maximum in the first half of the month of April (60 kg/m²) and at a minimum at the middle of the month (30 kg/m²). An attempt has been made to correlate the measured values of integrated water vapor content with those obtained from the radiosonde data, as shown in Figure 6.16. The regression analyses and the line between the radiometric and radiosonde estimations of vapor have been drawn as $V_{RADIOMETER} = a + bV_{RADIOSONDE}$ for three different frequency pairs: (1) 22.234 and 30 GHz pair, (2) 23.034 and 30 GHz pair, and (3) 23.834 and 30 GHz pair. While drawing the best-fit curve, we found that for the first pair, the line fits well with

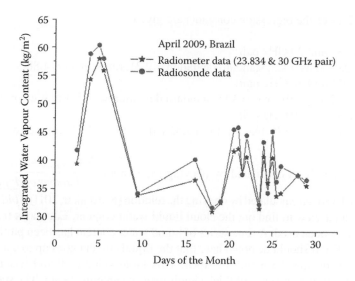

FIGURE 6.15 Time series of integrated water vapor content from radiometer data compared with radiosonde data.

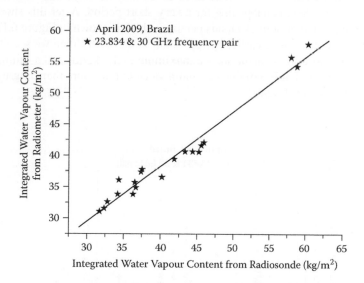

FIGURE 6.16 Correlations between measured and calculated values of vapor content using 23.834 and 30 GHz frequency pair.

$V_{RADIOMETER} = A \times V_{RADIOSONDE}$. The main reason for the choice of two frequency pairs is that the algorithm presented in this text is developed as a dual-frequency algorithm applicable in nonrainy conditions. Here, although the rainiest month, April 2009, has been chosen, care has been taken to exclude the rainy period data. This has been taken care of by observing the simultaneous data recorded by the co-located disdrometer.

The values of the regression constants are given below:

For 22.234 and 30 GHz pair:
  $a = 0.93$, bias = 4.3, standard deviation = 2.31, r.m.s. error = 4.88
For 23.034 and 30 GHz pair:
  $a = 1.27$, $b = 1.01$, bias = 3.14, standard deviation = 1.35, r.m.s. error = 3.42
For 23.834 and 30 GHz pair:
  $a = 3.195$, $b = 0.876$, bias = 2.15, standard deviation = 1.51, r.m.s. error = 2.64

The bias has been calculated by obeying the relation $\dfrac{\sum V_{radiometer} - V_{radiosonde}}{number\ of\ data}$, and
r.m.s. error has been calculated by obeying the relation (Karmakar, 2011) $\sqrt{bias^2 + sd^2}$.

Now with a view to find out the cloud liquid water content, Equations 6.25, 6.26, and 6.40 have been used. A time series of liquid water content has been plotted (refer to Figure 6.17). It should be noted here that the liquid content goes up to a maximum of 7.0 kg/m² during the month of April. This large value of liquid water content might be due to the presence of thick clouds over the antenna beam. The sudden fall of the graphical presentation is also observed at certain times. It might be due to the presence of thin cloud overhead. The thick cloud may be dispersed with time due to wind activity, which is happening for a very short period. After this short spell, again the accumulation of thick clouds persists, which is shown in Figure 6.17. If we look back to the work done earlier by Karmakar et al. (2001), then we see that over Kolkata the liquid water content takes a maximum of 1.7 kg/m² and a minimum of less than a kg/m². But at Brazil the maximum value attains more than 5 kg/m² and a minimum of less than 1 kg/m².

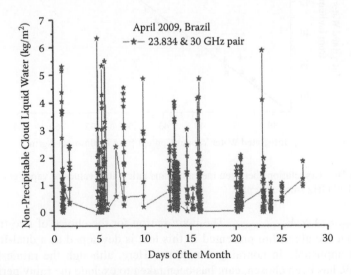

FIGURE 6.17 Time series of cloud liquid water content over Brazil using 23.834 and 30 GHz frequency pair.

The regression analysis between the estimated water vapor content by the radiometric method and the brightness temperature from the radiometer at 22.234 GHz shows the best linear equations over INPE, Brazil (22°S), as given by Karmakar et al. (2010).

$$W = 478.451T_b + 9574$$

A comparative study between the calculated (using the radiosonde data) values of water vapor content, y, and measured (using the radiometer) water vapor content, x, has been presented there as

$$y = 0.4218x + 31.31$$

Now taking advantage of latitudinal occupancy of the places of measurement, i.e., INPE, Brazil, and Institute of Radiophysics and Electronics, Kolkata, (22°N), India, we have presented the same regression analysis over Kolkata (Karmakar et al., 2001), found as

$$V_m \ (22.235 \ \text{GHz}) = 1.13646 \ V \ (\text{percentage error 5\%})$$

Here V_m is the measured water vapor content by using 22.235 GHz only and V is the calculated vapor content by using radiosonde data. But the use of dual-frequency measurement along with corresponding radiometric data analyses over Kolkata provides the relationship

$$V_m \ (22.235 \ \text{and} \ 31 \ \text{GHz}) = 1.0161 \ V \ (\text{percentage error 1.6062\%})$$

So it is worth noting that in estimating water vapor content by using the 22 GHz radiometer data only, the error is 5%, while that for measurement of vapor using both 22.235 and 31.4 GHz is 2% relative to the mean. However, the liquid water content over Kolkata, India (22°N), was observed to be as low as 0.02 kg/m² and as high as 1.85 kg/m².

REFERENCES

Adams, D.K., M.S. Rui, E.R. Fernandes, J.M. Maia, L.F. Sapucci, L.A.T. Machado, I. Vitorello, J.F.G. Monico, L.H. Kirk, S.I. Gutman, N. Filizola, and R.A. Bennett. A dense GNSS meteorological network for observing deep convection in the Amazon. *Atmos. Sci. Lett.*, 2011. doi: 10.1002/asl. 312.

Allnutt, J.E. Slant path attenuation and space diversity results using 11.6 GHz radiometer. *Proc. IEE*, 123, 1197–1200, 1976.

Battaglia, A., P. Saavedra, C. A. Morales and C. Simmer, Understanding 3D effects in polarized observations with the ground-based ADMIRARI radiometer during the CHUVA campaign, *Journal of Geophysical Research, Atmosphere*, 116:D09204, 2011, doi:10.1029/2010JD015335.

Bhattacharya, C.K. Microwave radiometric studies of atmospheric water vapour and attenuation measurements. PhD thesis, Benaras Hindu University, Varanasi, India, 1985.

Braunt, D. *Physical and dynamical meteorology*. Cambridge University Press, Cambridge, 1947.

Bohlander, R.A., R.J. Emery, D.T. Llewellyn-Jones, G.G. Gimmestad, O.A. Simpson, J.J. Gallagher, and S. Perkowitz. *Excess absorption by water vapour and comparison with theoretical dimer absorption in atmospheric water vapour*, ed. A. Deepak, T.D. Wilkersor, and L.H. Ruhnke. Academic Press, New York, 1980.

Cartelis, C., and C. Varotsos. Surface ozone in Athens, Greece, at the beginning and at the end of the 20th-century. *Atmos. Environ.*, 28, 3–8, 1994.

Cimini, D., E.R. Westwater, A.J. Gasiewski, M. Klein, V. Leuski, and J.C. Liljegren. Ground-based millimeter and submillimeter-wave observations of low vapor and liquid water contents. *IEEE Trans. Geosci. Remote Sensing*, 45, 2169–2180, 2007.

David, K. Adams, Rui, M.S., Fernandes, E.R., Maia, J.M., Sapucci, L.F., Machado, L.A.T., Icaro Vitorello, Jo~ao Francisco Galera Monico, Kirk, L.H., Gutman, S.I., Naziano Filizola, Richard, A. Bennett. A dense GNSS meteorological network for observing deep convection in the Amazon. Atmos. Sci. Lett. 312, 2011, doi:10.1002/asl. Published online in Wiley Online Library (http://onlinelibrary.wiley.com) (2011).

Gibbins, C.J. Improved algorithm for the determination of specific attenuation at sea level by dry air and vapour in the frequency range 1–350 GHz. *Radio Sci.*, 21, 949–954, 1986.

Grody, C. Remote sensing of the atmospheric water content from satellite using microwave radiometry. *IEEE Trans. Antennas Propagation*, 24, 155–162, 1976.

Grody, N.C., A. Gruber, and W.C. Shen. Atmospheric water content over the tropical Pacific derived from Nimbus-6 scanning microwave spectrometer. *J. Appl. Meteorol.*, 19(8), 968–996, 1980.

Haas, R., T. Ning, and G. Elgered. Observation of long term trends in the amount of atmospheric water vapor by space geodesy and remote sensing techniques. Presented at Geoscience and Remote Sensing Symposium (IGARSS), 2010 IEEE International, 2010, pp. 2944–2947.

Hess, L.T. *Introduction to theoretical meteorology*. Heidy Holt, New York, 1959.

Hocke, K., N. Kämpfer, C. Gerber, and C. Mätzler. A complete long-term series of integrated water vapour from ground-based microwave radiometers. *Int. J. Remote Sensing*, 32(3), 751–765, 2011.

Hogg, D.C., F.O. Guiraud, J.B. Sinder, M.T. Decker, and E.R. Westwater. A steerable dual-channel microwave for measurement of water vapor and liquid in the troposphere. *J. Climate Appl. Meteorol.*, 22, 789–806, 1983.

Janssen, M.A. A new instrument for the determination of radio path delay due to atmospheric water vapour. *IEEE Trans. Geosci. Remote Sensing*, 23, 455–490, 1985.

Karmakar, P.K. Studies of microwave and millimeter wave propagation. PhD thesis, University of Calcutta, India, 1989.

Karmakar, P.K. *Microwave propagation and remote sensing*. CRC Press, Taylor and Francis Group, Boca Raton, FL, 2011.

Karmakar, P.K., S. Chattopadhyay, and A.K. Sen. Estimates of water vapour absorption over Calcutta at 22.235 GHz. *Int. J. Remote Sensing*, 20, 2637–2651, 1999.

Karmakar, P.K., M. Maiti, A.P.J. Calheiros, C.F. Angelis, L.A.T. Machado, and S.S. Da Costa. Ground based single frequency microwave radiometric measurement of water vapour. *Int. J. Remote Sensing*, 2010. doi: 10.1080/01431161.2010.543185.

Karmakar, P.K., M. Maiti, S. Sett, C.F. Angelis, and L.A.T. Machado. Radiometric estimation of water vapour content over Brazil. *Adv. Space Res.*, 2011. doi: 10.1016/j.asr.2011.06.032.

Karmakar, P.K., M. Rahaman, and A.K. Sen. Measurement of atmospheric water vapour content over tropical location by dual frequency microwave radiometry. *Int. J. Remote Sensing*, 22, 3309–3322, 2001.

Kondratyev, K.Y., and C.A. Varotsos. Global tropospheric ozone dynamics. Part I. Tropospheric ozone precursors. *Environ. Sci. Pollut. Res.*, 8, 57–62, 2001a.

Kondratyev, K.Y., and C.A. Varotsos. Global tropospheric ozone dynamics. Part II. Numerical modeling of tropospheric ozone variability. *Environ. Sci. Pollut. Res.*, 8, 113–119, 2001b.

Liebe, H.J. MPM—An atmospheric millimeter wave propagation model. *Int. J. Infrared Millimeter Waves*, 10, 631–650, 1989.

Maitra, A., and K. Chakraborty. Cloud liquid water content and cloud attenuation studies with radiosonde data at a tropical location. *J. Infrared Millimeter Terahertz Waves*, 30, 367–373, 2009.

Moran, J.M., and B.R. Rosen. Estimation of propagation delay through troposphere and microwave radiometer data, *Radio Science*, 16, 235–244, 1981.

Pandey Prem, C., Gohil, B.S., Hariharan, T.A. A two frequency algorithm differential technique for retrieving precipitable water from satellite microwave radiometer (SAMIR-II) on board Bhaskara II. IEEE *Trans. Geo Sc. and Remote Sensing* 22, 647–655, 1984.

Raju C. Suresh, R. Renju, Tinu Antony, Nizy Mathew, and K. Krishna Moorthy, Microwave Radiometric Observation of a Waterspout Over Coastal Arabian Sea, IEEE Geoscience and Remote Sensing Letters, (accepted), 2013.

Resch, G.M. *Another look at the optimum frequencies for water vapour radiometer.* TDA Progress Report. TDA, 1983.

Salonen, E. New prediction method of cloud attenuation, *Electronics Letter.* 27, 1106–1108, 1991.

Sen A.K, Karmakar P.K, Das T.K, Dev Gupta A.K, Chakraborty P.K, Devbarman S. Significant heights for water vapour content in the atmosphere *Int. J. Remote Sensing*, (U.K.) vol. No. 10, pp: 1119–1124, 1989.

Sherwood, S.C., R. Roca, T.M. Weckwerth, and N.G. Andronova. Tropospheric water vapour convection and climate: A critical review. *Rev. Geophys.*, 48, RG2001, 2009. doi: 10.1029/2009RG000301.

Simpson, P.M., E.C. Brand, and C.L. Wrench. *Liquid water path algorithm development and accuracy. Microwave radiometer measurements at Chilbolton.* Radio Communications Research Unit CLRC—Rutherford Appleton Laboratory Chilton, DIDCOT, Oxon, UK, 2002.

Varotsos, C. Atmospheric pollution and remote sensing: Implications for the hemisphere ozone hole split in 2002 and the northern mid-latitude ozone trend. *Adv. Space Res.,* 33, 249–253, 2004.

Varotsos, C., Kondratyev, K.Y. and Efstathiou, M., 2001, On the seasonal variation of the surface ozone in Athens, Greece. *Atmos. Environ.*, 35, pp. 315–320.

Waters, J.W. Absorption and emission by atmospheric gases. In *Methods of experimental physics*, ed. M.L. Meeks. Academic Press, New York, 1976.

Westwater, E.R. Ground based determination of low altitude temperature profiles by microwaves. *Monsoon Weather Rev.*, 100, 15–18, 1972.

Westwater, E.R. The accuracy of water vapor and cloud liquid determination by dual-frequency ground-based microwave radiometry. *Radio Sci.*, 13, 677–685, 1978.

Westwater, E.R., and F.O. Guiraud. Ground-based microwave radiometric retrieval of precipitable water vapor in the presence of clouds with high liquid content. *Radio Sci.*, 13, 947–957, 1980.

7 Microwave Radiometric Estimation of Excess Electrical Path

7.1 INTRODUCTION

Electromagnetic waves are potentially used in geodetic measurements like distance, directions, delay, etc. Depending on wavelength and purposes, the measurements of the parameters, like amplitude, phase, and angle of arrival, are of paramount interest. However, the effect of earth's atmosphere offers a serious limitation to the accuracy and precision of geodetic measurements. Atmospheric water vapor is a limiting source of error (Resch, 1984) in determining the baselines by the technique of very long-baseline interferometry (VLBI). The use of a global positioning system (GPS) offers even more precision geodetic measurements (Bevis et al., 1992). On the other hand, it may be even more limited by the variable wet path delay component due to the very presence of water vapor. As early as 1976, a series of efforts began on the development of microwave remote sensing systems to determine the wet path delay for geodetic applications (Janssen, 1985). These efforts were culminated in 1982 with the asset of seven water vapor radiometers dedicated to the task of providing real-time path delay measurements at Goddard Space Flight Center (Resch, 1984).

In fact, the significance of atmospheric effects on the geodetic observations is clearly recognized by several scientists (Grafarend et al., 1978). Dealing properly with the atmospheric effects is definitely a key factor for operational geodesy. Basically, there are two different possible approaches: (1) the integral atmospheric model, where the atmospheric effects are incorporated in some form in the adjustment process, and (2) the peripheral atmospheric model, where the observations are corrected for the atmospheric effects before they are entered into the adjustment process. As such, no clear guideline is available so far for the decision of when to treat a correction on the geodetic observations by an integral or peripheral model. However, the peripheral model suffers from a shortcoming, as the observational errors are not separated from the model errors. On the other hand, the treatment of the atmospheric effects in an integral model appears to be better and yields information for the model development process.

The development of the integral model starts with the appropriate equations derivable from the geodetic measurements. Since the atmosphere affects the measurements, geodesists are able to estimate those properties to which their observations are sensitive. Among these, the vertically integrated delay of the radio signal due to water vapor in the atmosphere creates a lot of attraction. Because of the importance of water vapor to meteorology, the prospect of a new, relatively inexpensive instrument to determine its spatial and temporal distribution should be welcomed (Bevis et al., 1996). The operational tool for determining the distribution of water vapor in the atmosphere has been the radiosonde (Westwater, 1997). This provides vertical profile information about the meteorological variables, like pressure (p), temperature (T), and relative humidity (RH), but the operational cost restricts their use by the National Weather Service (NWS) and other national agencies to only twice per day. GPS, VLBI, and most water vapor radiometers (WVRs) that are in current use continuously measure the integrated properties of the atmosphere, in an unmanned way.

In the zenith direction a water vapor-induced path delay may vary from less than 1 to more than 30 cm, depending on prevailing climate and weather (Claflin et al., 1978). A WVR with temperature and water vapor profiling capability may provide a solution in this context (Solheim et al., 1998). Although it does not have the vertical resolution as those of radiosonde, it has the advantage of high temporal resolution and is able to detect cloud liquid water. Karmakar et al. (2002) shows the effect of liquid water in the millimeter-wave spectrum at a tropical location. However, the presence of liquid water cause a small change in path delay (Resch, 1984). But for the present purpose, we need to discriminate the effect of liquid, however small it is. It is always suggested (Karmakar et al., 2011) to use the resonant frequency and the other one at around atmospheric window frequency, provided the vertical profiles of pressure and temperature are constant (Resch, 1983). But this does not happen in practice. The choice of the resonant frequency is precluded because of its dependence on pressure broadening. Hence, it is suggested to use the slightly off-line frequency, which is not pressure dependent. Westwater (1978) showed that the frequencies independent of pressure lie both ways around the 22.234 GHz line. It is suggested to measure the propagation path delay over a tropical location by exploiting the radiometric measurement of the brightness temperatures at 23.834 and 30 GHz (Karmakar et al., 2011). The use of dual-frequency measurement will take care of nonprecipitating cloud liquid water (Simpson et al., 2002).

7.2 THE PROBLEM

The effects of the troposphere on the propagation of microwave signals are treated in detail by Bean and Dutton (1968), and thereafter reviewed by Thompson (1975) and Resch (1984). The apparent electric path L_e, along some atmospheric path L, is defined as

$$L_e = \int n(h)\,dh$$

where n is the refractive index at h. The excess path delay is defined as

$$\Delta L = L_e - L \tag{7.1}$$

and the refractivity N is

$$N = (n - 1) \times 10^6$$

to get

$$\Delta L = 10^{-6} \int N dh \qquad (7.2)$$

Here it may be noted that the total refractivity is a sum total of dry term $N_d = 77.6 P/T$ and wet term $N_V = (3.73 \times 10^5) e/T^2$. The dry delay mainly depends on the amount of air through which the signal propagates. Hence, it can be easily modeled with surface pressure measurements. The term *delay* refers to change in path length due to change in refractive index during the propagation of radio signals through the atmosphere duly constituted by several gases. Their combined refractive index is slightly greater than unity and gives rise to a decrease in signal velocity. This eventually increases the time taken for the signal to reach the receiving antenna (Adegoke and Onasanya, 2008). The wet path delay depends on the precipitable water vapor in the column of air through which the signal propagates. Several methods for the estimation of wet path delay have been suggested by Rocken et al. (1991). Besides this, bending of the ray path also increases the delay (Collins and Langley, 1998). The refractivity N is divided into two parts. The term N_h is the refractivity due to gases of air except water vapor and is called the hydrostatic refractivity. The term N_W is the refractivity due to water vapor and is called wet refractivity. Hence, from Equation 7.2 it can be written as

$$\Delta L = 10^{-6} \int N_h dh + 10^{-6} \int N_w dh \qquad (7.3)$$

According to Maiti et al. (2009), N_h and N_W yield both dry and wet path delay corrections and are given by

$$N_h = k_1 \left[\frac{P_d}{T} \right] \qquad (7.4)$$

$$N_w = k_2 \left[\frac{P_w}{T} \right] + k_3 \left[\frac{P_w}{T^2} \right] \qquad (7.5)$$

Here P_d and P_w are the partial pressure due to gases and water vapor pressure, respectively, and T is the ambient atmospheric temperature (K). The best average rather than best available coefficients provides a certain robustness against unmodeled systematic errors and increases the reliability of K values, particularly if data from different laboratories can be averaged. However, the best available coefficient, according to Rueger (2002), is given by

$$k_1 = 77.674 \pm 0.013 \; K/hPa$$

$$k_2 = 71.97 \pm 10.5 \; K/hPa$$

$$k_3 = 375406 \pm 3000 \; K^2/hPa$$

The bulk of delay correction is due to the dry component. Typically, the dry delay is of the order of 230 cm at the zenith. According to Hopfield (1971, 1976) the dry delay (cm) term at the zenith is given by

$$\Delta L_D = 0.2276\, P_s \qquad (7.6)$$

Here P_s is the surface pressure in mb (hectopascal) with an accuracy of better than 0.5 cm in the zenith direction. Gardner (1976) pointed out that at low elevation angles the r.m.s. delay error, although nonnegligible, is small compared to the error involved in estimating the wet delay from surface measurements. According to Hess (1959), the wet delay (cm) at the zenith is given by

$$\Delta L'_v = 1.73 \times 10^{-3} \int (\rho_v / T)\, dh \qquad (7.7)$$

Here ρ_v is the water vapor density (g/m^3), and h is in meters. Reber and Swope (1972) have shown that the correlation between the total precipitable water vapor and surface absolute humidity is highly variable. This suggests that the surface-based modeling of delay will not be so accurate, and that the effects of horizontal gradients in vapor distribution may be relatively larger than the gradients in the dry distribution. It is also suggested by Resch (1984) that only the passive microwave radiometry may provide an accuracy of ± 1 cm delay or even better at a chosen frequency pair, depending upon location and season. However, following Equation 7.7, one can measure the wet delay, but in that case the accuracy may go up to 2 to 5 cm. On the other hand, the radiometer will not provide the information about delay during raining conditions, but it would be very much useful in clear and cloudy skies, for which a proper algorithm has to be developed.

7.3 THEORETICAL MODEL

According to Chandrasekhar (1960), the absorption and emission of radiation from an atmosphere both are governed by the equation of radiative transfer. Considering the nonscattering medium, if we observe an elevated line of sight free from any obstruction and can neglect radiation from extraterrestrial sources, the radiative transfer equation takes the form

$$T_B = T_C e^{-\tau} + T_m (1 - e^{-\tau}) \qquad (7.8)$$

where $T_C = 2.9K$ is the cosmic background temperature, T_m is the frequency-dependent mean atmospheric temperature (Ulaby et al., 1986), and τ is the total atmospheric opacity τ, which is defined as

$$\tau = \int (\alpha_V + \alpha_L + \alpha_d)\, ds = \tau_V + \tau_L + \tau_d \qquad (7.9)$$

Here α's are the corresponding vapor, liquid, and dry absorption coefficients.

The radiation is a nonlinear function of the required quantities like absorption and emission, and we are to linearize the expression around a suitably chosen first guess, such as a climatological mean. If we look into Equation 7.8, we see that the exponential term will prevent the brightness temperature from increasing linearly with increasing path delay. It is clear that at a certain value of opacities at a particular frequency pair the brightness temperature will saturate and will be equal to surface temperature. Hence, it can be well recognized that under this situation the measurement will be completely insensitive to water vapor path delay. It is now becoming an essential need to overcome this situation. The solution is to linearize the brightness temperature by transforming the calculation of path delay into a problem of solving two linear equations with two unknown quantities. Here, they are (1) water vapor path delay and (2) integrated liquid water.

According to Wu (1978, 1979), the linearized brightness temperature is written as

$$T_B' = T_c - \{T_{eff}' - T_c\} \ln\left(1 - \frac{T_B - T_c}{T_{eff} - T_c}\right) \tag{7.10}$$

Here

$$T_{eff} = \frac{\int_0^\infty T\alpha e^{-\tau(s)} ds}{\int_0^\infty \alpha e^{-\tau(s)} ds} \tag{7.11}$$

$$T_{eff}' = \frac{\int_0^\infty T\alpha\, ds}{\int_0^\infty \alpha\, ds} \tag{7.12}$$

Figure 7.1 shows the linearization correction $T_B' - T_B$ as a function of T_B for three estimates of $T_{eff} = T_{eff}'$ (for small opacities). However, for very high brightness

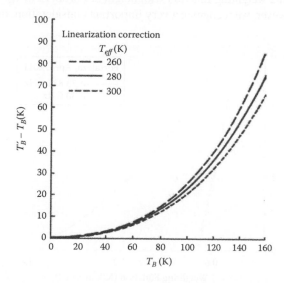

FIGURE 7.1   Correction $T_B'-T_B$ required to linearize brightness temperature T_B for three effective temperatures T_{eff}.

temperature, the correction becomes significant. It becomes highly necessary for a tropical zone but not for a temperate zone. This is because of large abundances of vapor content over a tropical zone compared to over a temperate zone. However, in the present context we shall confine ourselves with path delay only, leaving aside finding the liquid content.

We describe changes in the brightness temperature around the first guess by means of weighting functions W_ρ, W_T, and W_L (frequency-dependent term), which express the sensitivity of T_B to the variation of the humidity $\Delta\rho(h)$ or the temperature $\Delta T(h)$ around their initial values. Now, calling the other variables except temperature generic, we write the weighting functions, say, for water vapor, as

$$W_x = e^{-\tau(0,h)} \frac{\partial\alpha(h)}{\partial x}\left[T(h) - T_{bg}e^{-\tau(s,\infty)} - \int_h^\infty T(h')\alpha(h')e^{-\tau(h,h')}dh'\right] \quad (7.13)$$

To explain more clearly about the weighting function, we first take the temperature weighting function (km^{-1}). If we have a $\delta T(K)$ change in T over a height interval δh (km), the brightness temperature response $\delta T_b(K)$ to this change is $\overline{W}_T\delta T\delta h$, where $\overline{W}_T$ is called the height average of W_T over the height interval δh. For water vapor, similarly, $\delta T_b(K) = \overline{W}_\rho\delta_\rho\delta h$. If the units of ρ are $g\cdot m^{-3}$ and h is in km, then their product $\delta V = \delta\rho\cdot\delta h$ has the same unit of mm. Thus, $\delta T_b(K) = \overline{W}_\rho\delta V$ (Westwater et al., 1990). The weighting functions are determined from the height profile of attenuation coefficients at different frequencies. These in turn depend on the water vapor content at the place in question. The constancy of the weighting function with height at desired frequency provides application potential of that frequency. Figure 7.2 shows the variation of the weighting function with height at Fortaleza (3°S), Brazil, a tropical location. However, we recognize a very important consideration, that a frequency

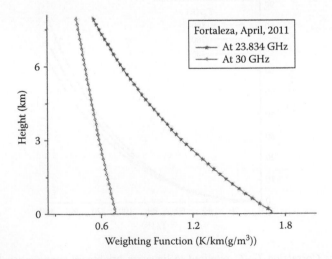

FIGURE 7.2   Water vapor weighting function at 23.834 and 30 GHz at Fotaleza (3°S), Brazil.

pair may be chosen to minimize variation of weighting function $W(s)$ with altitude (Wu, 1979). The combined weighting function is given by

$$W(h) = \frac{T(T - T_c)}{\rho} \left[\frac{\alpha_{v,1}}{f_1^2} - \frac{\alpha_{v,2}}{f_2^2} \right] \tag{7.14}$$

where α_v is the absorption coefficient for water vapor and ρ is the water vapor density. Care has to be taken to select the frequency pairs, particularly for the coastal regions, because layering of marine and continental air is a frequent occurrence.

Now if we assume the brightness temperature of the atmosphere T_B at two frequencies $f_1 = 23.834$ GHz and $f_2 = 30$ GHz, we write the linear equations

$$\tau_1 = W_{\rho 1} (\Delta L_V) + \tau_{L1} + \tau_{d1} \tag{7.15}$$

$$\tau_2 = W_{\rho 2} (\Delta L_V) + \tau_{L2} + \tau_{d2} \tag{7.16}$$

Here τ_1 = opacity (dB) at frequency $f_1 = 23.834$ GHz, τ_2 = opacity (dB) at frequency $f_2 = 30$ GHz, $W_{\rho 1}$ = water vapor weighting function at frequency $f_2 = 23.834$ GHz, $W_{\rho 2}$ = water vapor weighting function at frequency $f_2 = 30$ GHz, τ_{L1} = opacity due to liquid water at frequency $f_1 = 23.834$ GHz, τ_{L2} = opacity due to liquid water at frequency $f_2 = 30$ GHz, τ_{d1} = dry air opacity at frequency $f_1 = 23.834$ GHz, and τ_{d2} = dry air opacity at frequency $f_2 = 30$ GHz.

It is to be mentioned that the zenith opacity may be obtained by using the following relation (Allnutt, 1976):

$$\tau(dB) = 10 \ log_{10} \frac{T_m - T_C}{T_m - T_B} \tag{7.17}$$

Here excess path delay due to water vapor is $\Delta L_V = L_e - L$, assuming the apparent or electrical length L_e, along some atmospheric path L.

Various approaches have been reviewed by Berman (1976) in this regard. In general, these methods assume a standard functional form for the vertical distribution of ρ_V and T so that the integral in Equation 7.13 cannot be determined by surface measurements. But Reber and Swope (1972) have shown that the correlation between total precipitable water vapor and surface humidity is highly variable, depending on site and season. This strongly suggests that surface models are not accurate, and also that the effects of horizontal gradients in the vapor distribution may be relatively larger than those of the gradients in the dry distribution.

However, attenuations due to cloud liquid at the two frequencies are related as (Staelin, 1966; Karmakar et al., 1999, 2001)

$$\tau_{L2} = \left(\frac{f_2}{f_1} \right)^2 \tau_{L1} \tag{7.18}$$

The dry opacity terms are related according to the relation (Resch, 1984)

$$\tau_{d2} = \beta \tau_{d1} \tag{7.19}$$

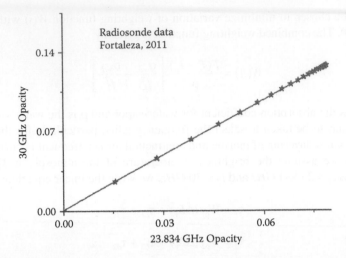

FIGURE 7.3 Dry opacity for 23.834 and 30 GHz at Fotaleza, Brazil.

where β depends on the pair of chosen frequencies (here $f_1 = 23.834$ GHz and $f_2 = 30$ GHz). The value of $\beta = 1.6047$ was evaluated at Fortaleza (3°S) (see Figure 7.3) for this particular pair of frequencies by using the MPM model (Liebe, 1989). It should be mentioned that the radiosonde data were usually collected by launching a balloon containing sensors for measuring atmospheric parameters such as temperature, pressure, etc., at different vertical levels. These data were then transmitted to a local station, which is a few meters away from the site where the radiometer is located. It should be noted that the balloon during its ascent would follow the wind field independently, which may affect the baseline between radiometer and balloon. Now using Equations 7.18 and 7.19, we write Equation 7.12 as

$$\tau_2 = W_{\rho 2}(\Delta L_V) + \left(\frac{f_2}{f_1}\right)^2 \tau_{L1} + \beta \tau_{d1} \tag{7.20}$$

and solve to give the excess path delay due to water vapor:

$$\Delta L_V = \left\{ W_{\rho 1}\left(\frac{f_2}{f_1}\right)^2 - W_{\rho 2} \right\}^{-1} \left[\left\{ \left(\frac{f_2}{f_1}\right)^2 \tau_1 - \tau_2 \right\} + \tau_{d1}\left\{ \beta - \left(\frac{f_2}{f_1}\right)^2 \right\} \right] \tag{7.21}$$

Equation 7.21 provides the formal solution for the excess path delay due to water vapor in terms of the transformed observables τ_1 and τ_2.

But on the other hand, considering the combined weighting function as described in Equation 7.14, Wu (1979) has developed a method for finding ΔL_V, for a dual-channel radiometer, which is described as

$$\Delta L_v = (b_0' + b_1 T_{b,1}' + b_2 T_{b,2}') / W_0$$

where

$$b_0' = W_\rho b_0 - T_c(b_1 + b_2)$$

$$W_\rho = m(P/\bar{P})^2 \left(\frac{\bar{T}}{T}\right)^{1.85}$$

and W_0 is the combined weighting function at the ground.

For convenience of application, the above set of equations can be written as

$$\Delta L_v = a_0'm - T_c' + a_1 T_{B,1}' + a_2 T_{B,2}'$$

$$a_0' = \frac{b_0\left(\dfrac{P}{\bar{P}}\right)^2\left(\dfrac{\bar{T}}{T}\right)^{1.85}}{W'(0)}$$

$$T_c' = \frac{T_c(b_1 + b_2)}{W'(0)}$$

$$a_1 = \frac{b_1}{W'(0)}$$

$$a_2 = \frac{b_2}{W'(0)}$$

Here b_0, b_1, b_2, are the constants, P is the surface pressure in N/m^2, T is the surface temperature, and $T_{B,1}'$ and $T_{B,2}'$ are to be evaluated from measured brightness temperature at the two desired frequencies.

Elgered et al. (1991) adopted another model in which the zenith hydrostatic delay (ZHD), in millimeters, is given by

$$\Delta L_H = ZHD = (2.2779 \pm 0.0024)P_s/f(\lambda, H)$$

where P is the total pressure (in millibars) at the earth's surface, and

$$f(\lambda, H) = (1 - 0.00266 \, cos2\lambda - 0.00028H)$$

accounts for the variation in gravitational acceleration with latitude λ and the height H of the surface above the ellipsoid (in kilometers). The troposphere accounts for about 75% of the total hydrostatic delay.

The second component of total delay is known as the wet delay, ΔL_W. The zenith wet delay (ZWD) is given by

$$\Delta L_W = ZWD = 10^{-6} \left[K_2' \int \left(P_v / T \right) dh + K_3 \int \left(P_v / T^2 \right) dh \right]$$

where $K_2' = (17 \pm 10) \, mbar^{-1}$, the integral is along the zenith path, and the delay is given in units of h (Davis et al., 1985). It is usually adequate to approximate this expression as (Bevis et al., 1992)

$$\Delta L_W = (0.302 \pm 0.004) \, K^2 mbar^{-1} \int \left(P_v / T^2 \right) dh$$

These expressions can be evaluated from profiles of P_v and T provided by radio-sondes, if such data are available. Almost all of the wet delay occurs in the troposphere, and most of it occurs in the lower troposphere.

7.4   DETERMINATION OF CONSTANTS IN THE ALGORITHM

In the previous sections, while deriving the ΔL_v, we needed to use terms like $W_{\rho 1}$, $W_{\rho 2}$, T_m, τ, and others. We also assumed these are either constants or, in the case of dry opacity, calculable from first principles. But in practice, we will find these are not actually constants, but instead show a clear seasonal variation, depending on sites. However, these variations are not large enough to invalidate our underlying assumption, but they do indicate that we must consider the constants in our algorithm as a variable for each individual site. These are to be found by some proper statistical regression analyses depending on the local radiosonde data. This database will then be used to determine the value of the constants in evaluating the ΔL_v, as mentioned in Equation 7.21. The variable W_ρ is to adjust for changes in air mass, m, and for changes in the oxygen absorption with change in surface pressure and temperature. $\overline{T}$ and $\overline{P}$ are surface values and are best chosen representative of the operating conditions. The water vapor absorption coefficient α_v may be calculated by using the well-known MPM model, as discussed earlier. The coefficients b_0, b_1, b_2 may be determined from the regression of measured brightness temperature against path delay from simultaneous radiosonde data. However, a_0', T_c', a_1, a_2 (see the algorithm as described by Wu) are to be prepared for suitable choice of P, T, and ρ values.

Initially, it is to be confirmed that the radiosonde launch and the radiometer are collocated. Now let us suppose that several launches have been made from the same site of the radiometer, and then we minimize (see Equations 7.17 and 7.21)

$$S = \sum \left(\Delta L_v - \Delta L_v' \right)^2 \tag{7.22}$$

with respect to b_0, b_1, b_2 with several constraints. The first constraint is eliminated by using the following equation, similar to Equation 7.18:

$$\frac{b_1}{b_2} = -\left(f_2 / f_1 \right)^2 \tag{7.23}$$

The other constraints demand that the best-fitted line between the measured and calculated path delay should have unity slope and zero intercept by adjusting the radiosonde data profile, i.e., $\sum L_{v,i} = \sum L'_{v,i}$. This can be done by minimizing the values of b_1 and b_2, which are too small and offset by a value of b_0.

Coefficients may be calculated from measured brightness temperature data using a constrained regression analysis against a radiosonde-measured path delay or a constrained regression analysis of theoretically calculated brightness temperature against a radiosonde-measured path delay.

Since our constants are not really constants, we must expect some level of algorithm noise. We can best evaluate this noise level by determining the constants in a regression analysis. In this case, we shall use the radiosonde data to compute path delay and solve the equation of radiative transfer for the associated brightness temperature at the selected frequencies. This database will be used to minimize the level of residuals by least square sense. Still, we will be left with the problem of relating the brightness temperature scale of the radiometer to the temperature scale as defined in the radiative transfer equation. Since our knowledge of absorption coefficients/brightness temperature and calibration of the radiometer is not perfect, we must accept the fact that ultimately the radiometer data must be compared directly with some independent method of measuring path delay.

7.5 MEAN ATMOSPHERIC TEMPERATURE T_m AT MICROWAVE FREQUENCIES

We have already discussed how the radiometric frequency for the measurement of brightness temperature at the particular frequency is to be chosen for desired purposes. However, the evaluation of desired accuracy in radiometric measurements depends on several factors, such as the spatial and temporal humidity profile available from radiosonde data, which is not exact because the ascending balloon follows the random path for wind during telemetry. Besides these, the other source of error in brightness temperature measurement by a passive radiometer may creep in through the choice of mean atmospheric temperature. This is supposed to be the only parameter responsible in converting the brightness temperature T_b to the attenuation values. As expected, T_m is a function of frequency (Mitra et al., 2000) and three basic meteorological parameters: atmospheric pressure, temperature, and dew point temperature. T_d has a significant role in VHF and UHF bands, but in the microwave and millimeter-wave band, the variation also has a significant role. Karmakar (2011) reported that T_m remains more or less constant during the monsoon period over a tropical location. But in the nonmonsoon period T_m takes the highest value 295.5 K at 125 GHz and the lowest value of about 279 K at 22.235 GHz. Hence, we have taken the liberty to select a constant value of $T_m = 275K$ because of the fact that we are using the brightness temperature data for the monsoon period (December–April 2009) at Fortaleza, Brazil. According to Ulaby et al. (1986), T_m is usually estimated from the surface temperatures. Based on 24 radiosonde profiles, Wu (1979) developed a simple relation:

$$T_m = aT_g$$

It is found there that $a = 0.94$ for 30 GHz and $a = 0.95$ for frequencies between 20 and 24.5 GHz. But, Mitra et al. (2000) reported that over a tropical location $a = 0.86$ for 30 GHz. According to Resch (1984), the mean radiating temperatures were found to be 274.3K (approximately) for 20.7 GHz and 271.5 K (approximately) for 31 GHz, and a simple relation with surface temperature at 20.7 GHz as

$$T_m = 50.3 + 0.786 T_g$$

However, if the surface data are available, then it is always instructive to find the mean radiating temperature properly as a function of frequency at the place of choice.

7.6 RADIOMETRIC ESTIMATION OF DELAY OVER TEMPERATE LOCATIONS

Resch (1984) presented the summary of best-fit parameters (using Equation 7.22) for the water vapor algorithm involving brightness temperature at some places of temperate locations (see Tables 7.1 and 7.2) for a 20.7 and 31.4 GHz pair. The sites were: Portland, Maine, Pittsburgh, Pennsylvania, El Paso, Texas, and Oakland, California. The best-fit equation with no liquid was (with no measurement noise)

$$\Delta L_V = a_0 + a_1 \{T_1 - 0.4346T_2\}$$

The presence of cloud, with reference to Tables 7.1 and 7.2, is checked through radiosonde data by an indication of relative humidity that goes greater than 95%. This

TABLE 7.1

Best-Fit Parameters Involving Brightness Temperature: Cloud Model = 0 (no liquid): $\Delta L_v = a_0 + a_1(T_1 - 0.4346T_2)$

Site	$-a_0$	$-\sigma$	a_1	σ	r.m.s. (cm)
Portland	1.57	0.05	0.662	0.003	0.28
Pittsburgh	1.45	0.05	0.649	0.003	0.25
El Paso	1.39	0.06	0.610	0.004	0.26
San Diego	1.71	0.13	0.642	0.007	0.38
Oakland	1.63	0.13	0.644	0.007	0.35
All sites	1.62	0.05	0.646	0.002	0.41

TABLE 7.2

Cloud Model Statistical Parameters

Cloud Model	Average Residual	r.m.s.
1	0.29	0.45
2	0.20	0.58
3	-0.26	1.55

TABLE 7.3

Best-Fit Parameters Involving Opacity: $\Delta L_v = a_0 + a_1(\tau_1 - 0.4346\tau_2)$

Site (cm)	$-a_0$	σ	a_1	σ	r.m.s.
Portland	−0.01	0.04	160.5	0.6	0.24
Pittsburgh	−0.08	0.05	158.3	0.7	0.24
El Paso	0.03	0.05	151.4	0.9	0.25
San Diego	0.07	0.11	156.9	0.6	0.37
Oakland	0.07	0.10	158.6	0.7	0.32
All sites	0.07	0.03	157.9	0.5	0.36

TABLE 7.4

Cloud Models of Varying Thicknesses and Locations based on Table 7.3

Cloud Model	Average Residual	r.m.s.
1	0.18	0.35
2	0.17	0.35
3	0.14	1.37

indicates an equilibrium condition with cloud liquid. The relative humidity below 94% indicates the top and bottom of the cloud, and the altitude of these points is calculated by simple interpolation. Given the cloud thickness and altitude, the three models for cloud liquid density are given by Decker et al. (1978). In this respect, the model proposed by Salonen (1991) may be used, which is largely described by Karmakar (2011). Tables 7.3 and 7.4 summarize the best-fit parameters for the path delay, and the corresponding equation over the same places is

$$\Delta L_v = a_0 + a_1 \{\tau_1 - 0.4346\tau_2\}$$

Now comparing Tables 7.1 and 7.2 and Tables 7.3 and 7.4, we see that there is a considerable improvement with fitting while considering opacities. This algorithm is mainly suited when the surface data are not readily available.

7.6.1 Test for Instrument Stability

The best method to check the stability, as suggested by Resch, is to use two radiometers side by side observing the same portion of sky. The target temperature is variable, but if the radiometers are really observing the same solid angle, then any difference between the two outputs must have its origin within the instrument. If there is any difference between the two, then one really does not know which is correct. Figure 7.4 shows the result of the worst-case side-by-side test for 24 hours in Pasadena. The r.m.s. of the differenced path delay was 0.59 cm over this interval, and the average offset was −0.07 cm.

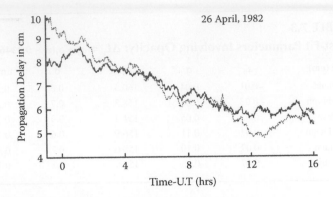

FIGURE 7.4   Side-by-side comparison of two zenith-looking radiometers.

However, there are also some drawbacks of this method. Since the radiometers are subjected to the same environment, it is possible that both instruments share some kind of systematic error that cancels in a common manner. But the long-term experimental observations may preclude this anomaly.

7.7   RADIOMETRIC ESTIMATION OF DELAY OVER TROPICAL LOCATION

The dual-frequency radiometer at 23.834 and 30 GHz was deployed at Brazil for the estimation of path delay. In this case Equation 7.22 was used. The radiometer outputs were taken in the form of sky brightness temperature (T_b). Radiometric opacities derived by using Equation 7.17 were used. For this purpose, the proper calibration method between the cold source (built into the radiometer) and sky brightness temperature was adopted.

The radiometric data were taken every 2 minutes. It should be mentioned here that January–February are the rainiest months of the year in Brazil. In spite of that, care has been taken to gather the data in no-rain conditions. This may help us to estimate the water vapor content and its corresponding attenuation, and hence the excess electrical path length in the rainiest month in Fortaleza (3°S), Brazil. The latitudinal occupancy for the measurement of water vapor over Brazil draws special attention because of entirely different environmental conditions, especially due to the Amazon Basin. For the sake of illustration, the diurnal variation of attenuation at 23.834 and 30 GHz at Fortaleza, Brazil, is presented in Figure 7.5.

Propagation delay values were calculated for the month of February (rainy season in the southern hemisphere) during the morning, afternoon, and evening spells (see Figure 7.6). These show that in the morning and evening spells the delay takes a maximum value of about 3 cm and a minimum of 0.85 cm. Figure 7.7 shows diurnal variation of delay for the whole month of February over Fortaleza, Brazil, taking into consideration that no rain data have been included. In this context it may be mentioned that Maiti et al. (2009) reported on the propagation delay over tropical locations using the refractivity data derived from the radiosonde data.

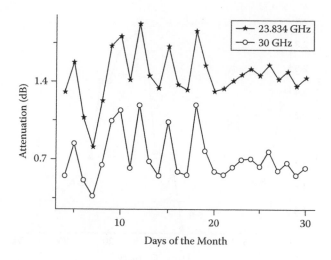

FIGURE 7.5   The diurnal variation of attenuation at 23.834 and 30 GHz at Fortaleza.

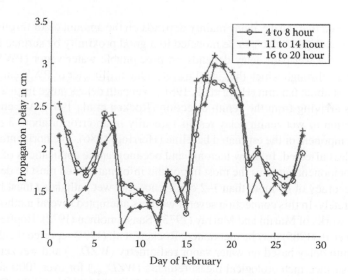

FIGURE 7.6   A plot for the propagation delay in the different spells (0004–0008, 0011–0014, and 0016–0020 hours) during February at Fortaleza. Median values were taken for the purpose.

7.8   VAPOR EFFECT ON BASELINE DETERMINATION

Space geodesy uses the difference in microwave signal arrival times at two sites to determine very accurate baseline vectors between these sites. The microwave signals originate from satellites (GPS) or quasars, and they experience propagation delays as they pass through the atmosphere, mainly through the troposphere. It has already

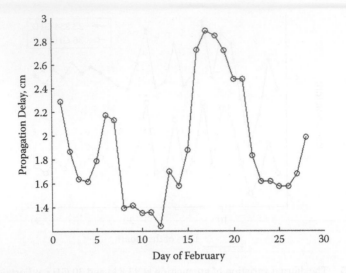

FIGURE 7.7 Tropospheric propagation delay for the entire month of February at Fortaleza.

been discussed that the dry delay mainly depends on the amount of air through which the signal travels, and also can be modeled to a good proximity by surface measurements. But, the wet path delay depends on precipitable water vapor (PWV) in the column of air through which the signal travels. One millimeter of PWV causes a wet path delay of about 6.5 mm (Hogg et al., 1981). Wet path delays range from 2 to 30 cm for signals arriving from the zenith direction (Rocken et al., 1991). A 1 cm error in the estimation of wet zenith delay results typically in an error of about 3 cm in the vertical component of the estimated baseline (Herring, 1986). The horizontal components are less affected. In many tectonic and oceanographic applications of GPS, the vertical component contains the most important information and must be determined with an accuracy of no more than 1–2 cm. Hence, the wet path delay must be known very accurately. In this connection several workers attempted several methods, out of which the works of Marini and Murray (1973), Saastamoinen (1972), Hopfield (1969), and Davis et al. (1985) may be mentioned. Rocken et al. (1991) reported the difference in wet zenith delay based on water vapor radiometry (WZD_{wvr}) and wet zenith delay based on surface meteorological measurements (WZD_{smm}) for over 1000 simultaneous observations at five locations in the United States, Mexico, and the Caribbean. Assuming that the WZD_{wvr} is accurate, the r.m.s. error of WZD_{smm} was computed by fitting a straight line to these differences. The corresponding relationship was

$$RMS\ error = RMS(WZD_{WVR} - WZD_{smm})$$

$$= [0.08\ WZD_{smm} + 1.25]\ cm$$

For convenience, let us take an example. If WZD_{SMM} is computed to be 25 cm, then this correction will be, on average, 3.25 cm in the zenith direction. The error then may result in a vertical baseline as much as 9 cm.

Water vapor, although a minor constituent in the atmosphere, is a source of error in geodetic applications. This error can be reduced by using the radiometric passive sensing of water vapor. But at the same time, care has to be taken so that unwanted contributions of nonprecipitable cloud liquid water can be separated from true water vapor budget determination. This can only be done by deploying the dual-frequency microwave radiometric technique.

The measurement in terms of brightness temperature needs a proper calibration of the radiometer. The derived opacity may be used for deducing the excess path delay. In addition, the algorithm noise has to be minimized.

REFERENCES

Adegoke, A.S., and M.A. Onasanya. Effect of propagation delay on signal transmission. *Pac. J. Sci. Technol.*, 9(1), 13–19, 2008.

Allnutt, J.E. Slant path attenuation and space diversity results using 11.6 GHz radiometer. *Proc. IEE*, 123, 1197–1200, 1976.

Bean, B.R., and E.J. Dutton. *Radio meteorology*. Dover Publication, New York, 1968.

Berman, A.L. *The prediction of zenith range refraction from surface measurements*. JPL Technical Report 32-1602:1–40. JPL, Pasadena, CA, 1976.

Bevis, M., S. Businger, T.A. Herring, C. Rocken, and R.A. Anthes. The promise of GPS in atmospheric monitoring. *Bull. Am. Meteorol. Soc.*, 77, 5–18, 1996.

Bevis, M., S. Businger, T. Herring, C. Rocken, R. Anthes, and R. Ware. GPS Meteorology: Remote sensing of atmospheric water vapor using the global positioning system. *J. Geophys. Res.*, 97(d14), 15787–15801, 1992.

Chandrasekhar, S. *Radiative Transfer*. Dover Publications, New York, 1960.

Claflin, E.S., S.C. Wu, and G.M. Resch. *Microwave radiometer measurement of water vapour path delay: Data reduction techniques*. DSN Progress Report 42–48:22–27. Jet Propulsion Laboratory, Pasadena, CA, 1978.

Collins, P., and R.B Langley. Tropospheric propagation delay. How bad can it be? Presented at the ION GPS-98, 11th International Technical Meeting of the Satellite Division of ION, Nashville, TN, 1998.

Davis, J.L., T.A. Herring, I.I. Shapiro, A.E.E. Rogers, and E. Elegard. Geodesy by radio interferometry: Effects of atmospheric modeling errors on estimates of baseline length. *Radio Sci.*, 20, 1593–1607, 1985.

Decker, M.T., E.R. Westwater, and F.O. Guiraud. Experimental evaluation of ground based remote sensing of atmospheric temperature and water vapor profiles. *J. Appl. Meteorol.*, 17, 1788–1795, 1978.

Elgered, G., J.L. Davis, T.A. Herring, and I.I. Shapiro. Geodesy by radio interferometry: Water vapor radiometry for estimation of the wet delay. *J. Geophys. Res.*, 96, 6541–6555, 1991.

Gardner, C.S. Effects of horizontal refractivity gradients on the accuracy of laser ranging to satellites. *Radio Sci.*, 11, 1037–1044, 1976.

Grafarend, E.W., I.I. Mueller, H.B. Papo, and B. Richter. *Investigations on the hierarchy of reference frames in geodesy and geodynamics*, 127. NASA-CE-162499, HC A07/MF A01 CSCL 08G unclassified, G3/46 45195. Department of Geodetic Science, Ohio State University Research Foundation, 1978.

Herring, T.A. Precision of vertical position estimates from very long base line interferometry. *J. Geophys. Res.*, 91, 9177–9182, 1986.

Hess, L.T. *Introduction to theoretical meteorology*. New York, Heidy Holt, 1959.

Hogg, D.C., F.O. Guiraud, and M.T. Decker. Measurement of excess radio transmission length on earth space paths. *Astron. Astrophys.*, 95, 304–307, 1981.

Hopfield, H.S. Two-quatric tropospheric refractivity profile for correcting satellite data. *J. Geophys. Res.*, 74, 4478–4499, 1969.

Hopfield, H.S. Tropospheric effect on electromagnetically measured range: Prediction from surface weather data. *Radio Sci.*, 6, 357–367, 1971.

Hopfield, H.S. *Tropospheric effect on signals at very low elevation angle.* APL/JHU Tech Memo TG1291:1–39. Bethesda, MD 1976.

Janssen, M.A. A new instrument for the determination of radio path delay due to atmospheric water vapour. *IEEE Trans. Geosci. Remote Sensing*, GE-23(4), 485–490, 1985.

Karmakar, P.K. *Microwave propagation and remote sensing: Atmospheric influences with models and applications.* CRC Press, Boca Raton, FL, 2011.

Karmakar, P.K., S. Chattopadhyay, and A.K. Sen. Estimates of water vapour absorption over Calcutta at 22.235 GHz. *Int. J. Remote Sensing*, 20, 2637–2651, 1999.

Karmakar, P.K., M. Maiti, S. Chattopadhyay, and M. Rahama. Effect of water vapour and liquid water on microwave absorption spectra and its application. *Radio Sci. Bull.*, 303, 32–36, 2002.

Karmakar, P.K., M. Maiti, S. Sett, C.F. Angelis, and L.A.T. Machado. Radiometric Estimation of water vapour content over Brazil. *Adv. Space Res.*, 48, 1506–1514, 2011.

Karmakar P.K., M. Rahaman, S. Chattopadhyay, and A.K. Sen. Estimation of absorption and hence electrical path length at 22.235 GHz. *Indian J. Radio Space Phys.*, 30, 36–42, 2001.

Liebe, H.J. MPM—An atmospheric millimeter wave propagation model. *Int. J. Infrared Millimeter Waves*, 10, 631–650, 1989.

Maiti, M., A.K. Datta, and P.K. Karmakar. Effect of climatological parameters on propagation delay through the atmosphere. *Pac. J. Sci. Technol.*, 10(2), 14–20, 2009.

Marini, J.W., and C.W. Murray Jr. *Correction of lapse range tracking data for atmospheric refraction at elevations above 10 degrees.* Technical Memo TM-X-70555. NASA, Washington, DC, 1973.

Mitra, A., P.K. Karmakar, and A.K. Sen. A fresh consideration for evaluating mean atmospheric temperature. *Indian J. Phys.*, 74b(5), 379, 2000.

Reber, E.E., and J.R. Swope. On the correlation of the total precipitable water in a vertical column and absolute humidity. *J. Appl. Meteorol.*, 11, 1322–1325, 1972.

Resch, G.M. *Another look at the optimum frequencies for water vapour radiometer.* TDA progress report. TDA, 1983.

Resch, G.M. Water vapor radiometry in geodetic applications. In *Geodetic refraction*, ed. F.K. Burner, 53–84. Springer-Verlag, Berlin, 1984.

Rocken, C., J.M. Johnson, R.E. Neilan, M. Cerezo, J.R. Jordan, M.J. Falls, L.D. Nelson, R.H. Ware, and M. Hayes. The measurement of atmospheric water vapor: Radiometer comparison and spatial variation. *IEEE Trans. Geosci. Remote Sensing*, 29(1), 1991.

Rueger, M.J. Refractive index formulae for radio waves. Presented at FIG XXII International Congress, Washington, DC, 2002.

Saastamoinen, J. Contributions to theory of atmospheric refraction. Part 1. *Bull. Geodesique*, 105, 279–298, 1972.

Salonen, E. New prediction method of cloud attenuation. *Electronics Lett.*, 27(12), 1106–1108, 1991.

Simpson, P.M., E.C. Brand, and C.L. Wrench. *Liquid water path algorithm development and accuracy. Microwave radiometer measurements at Chilbolton.* Radio Communications Research Unit, CLRC—Rutherford Appleton Laboratory, Chilton, DIDCOT, Oxon, UK, 2002.

Solheim, F., J. Godwin, and R. Ware. Passive ground based remote sensing of atmospheric temperature, water vapor, and cloud liquid water profiles by a frequency synthesized microwave radiometer. *Meteorol. Z.*, 7, 370–376, 1998.

Staelin, D.H. Measurements and interpretation of the microwave spectrum of the terrestrial atmosphere near 1-cm wavelength. *J. Geophys. Res.*, 71(12), 2875–2882, 1966.

Thompson, M.C. Effects of the troposphere on the propagation time of microwave signals. *Radio Sci.*, 7, 727–733, 1975.

Ulaby, F.T., R.K. Moore, and A.K. Fung. *Microwave remote sensing: Active and passive*, 1. Artech House, Norwood, MA, 1986.

Westwater, E.R. The accuracy of water vapor and cloud liquid determination by dual-frequency ground-based microwave radiometry. *Radio Sci.*, 13, 677–685, 1978.

Westwater, E.R. Remote sensing of tropospheric temperature and water vapor by integrated observing systems. *Bull. Am. Meteorol. Soc.*, 78, 1991–2006, 1997.

Westwater, E.R., J.B. Sinder, and M.J. Fall. Ground based radiometric observation of atmospheric emission and attenuation at 20.6, 31.65, 90.0 GHz: A comparison of measurements and theory. *IEEE Trans. Antenna Propagation*, AP-38(10), 1569–1580, 1990.

Wu, S.C. Microwave radiometer measurement of water vapor path delay. DSN Progress Report 42–48. Jet Propulsion Laboratory, Pasadena, CA, 1978.

Wu, S.C. Optimum frequencies of a passive radiometer for tropospheric path length correction. *IEEE Trans. Antenna Propagation*, AP-27, 233–239, 1979.

8 Characterization of Rain and Attenuation in the Earth–Space Path

8.1 INTRODUCTION

The increasing demand for new communication channels, with higher data rates and thus wider bandwidths, imposes huge demands on the electromagnetic spectrum. Now, in order to accommodate and facilitate the development of communication systems, it becomes necessary to consider exploitation and utilization of higher and higher frequencies, extending well into the microwave and millimeter-wave spectrum. These regions of the electromagnetic spectrum are now becoming more accessible through recent advances in component and system technology. This in turn improves the availability of cost-effective, reliable, and compact hardware, thus creating new opportunities and possibilities hitherto either not achievable or not practicable at lower frequencies.

However, at microwave, in fact, above about 10 GHz, rain is a dominant source of signal attenuation in a communication link either in earth-space or in a line-of-sight path. This initiates more and more information on the interaction between microwave and rain, in order to develop reliable and accurate propagation models and prediction procedures. Of primary importance to the application of such prediction techniques are the statistical data on the distribution of rain intensities, on the duration of rain events with given intensities, and on the return period between such events. The propagation effects over the earth-space path at these bands are in the form of attenuation, depolarization, and scintillation, which are mainly caused by rain (Chakravarty and Maitra, 2010). Most of the attenuation measurements carried out at the temperate regions account for the stratiform nature of rain with relatively large rain cell diameters prevailing over a longer time with uniform light showers (Ramchandran and Kumar, 2006). But, on the other hand, for the tropical region, the situation is somewhat different. In this case, convective rainfall with heavy showers for a shorter period accompanies the stratiform nature of rain, thus playing an important role of variation of climatic behavior for these regions. Besides these, it has been reported by Ramchandran and Kumar (2006), Pan et al. (2001), and Bowthrope et al. (1990) that the attenuation increases with a decrease in elevation angle in the temperate region due to large cell size, whereas in the tropics the situation is reverse.

Variability of raindrop size distribution (DSD) in different climatic zones, particularly in the tropical zones, has also been a major concern in developing the necessary models applicable to everywhere over the globe. But the inadequacy of the experimental observations over most of the tropical zones is actually becoming a prime factor in developing a model on a global basis. However, to get a model of rain attenuation on the basis of available DSD over tropical zones, it is becoming important to assess the variability of attenuation that can occur due to differences in DSD in a particular region.

8.2 RAIN RATES, DURATION, AND RETURN PERIODS

The rain rate profiles during the year 1985–1986 were recorded at the Institute of Radio Physics and Electronics, University of Calcutta, Kolkata (22°N), India, using a fast-response rain gauge with an integration time of 14.3 seconds. A histogram of rain rate during that period, which is comparable to a similar histogram for a longer period of 4 years obtained from the measured rain rate in the UK (Harden et al., 1977), is shown in Figure 8.1(a, b). To elaborate our study, we present here histograms over Cachoeira Paulista (CP) (22°S), during 2009, and Fortaleza (3°S), Brazil, for a period of 2004–2007 (Figure 8.1(c, d)). The distribution of duration of rain in Kolkata and the UK is presented in Figure 8.2. This exhibits a similar trend, except a few cases of very high rain rates occurring in Kolkata. On the other hand, during the period (2004–2007) over Fortaleza, the total number of events was more than 900 up to 2 mm rain accumulation, but this number drastically goes down to 300 for rain

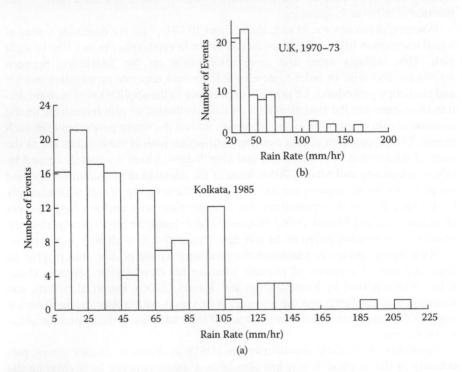

FIGURE 8.1 (a, b) Histogram of rain intensity at Kolkata, India, and the UK. (c, d) Histogram of rain intensity at Cachoeira Paulista and Fortaleza, Brazil.

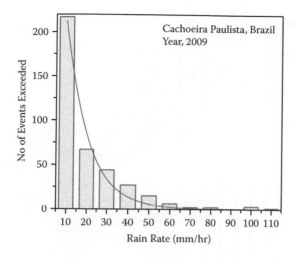

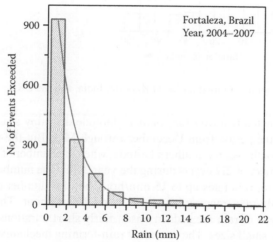

FIGURE 8.1 (Continued)

accumulation between 2 and 4 mm. This shows the large abundances of rain events over a tropical location like CP or Fortaleza. The histogram over both places shows exponential decay distribution with decay constants of 10.85 and 2.07 over CP and Fortaleza, respectively. The disdrometer used at CP is a Parsivel laser-based optical system for measurement of all types of precipitation. The size range of measurable liquid precipitation particles is from 0.2 up to 5 mm, and for solid precipitation particles it is from 0.2 up to 25 mm. In the process, precipitation particles can have a velocity of 0.2 up to 20 m/second. The rain monitor has the capability of measuring maximum rain rate up to 250 mm/hour and a minimum up to 0.0005 mm/hour with a resolution of 10 seconds. The maximum rain rate observed over the observational site was 107 mm/hour, and the minimum was less than 1 mm/hour. It is also interesting to note that only on two occasions did the rain rate go beyond 100 mm/hour.

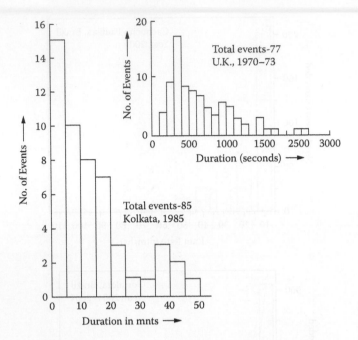

FIGURE 8.2   Histogram of rain duration at Kolkata, India, and the UK.

It should also be noted that over the southern latitude the rainy season is normally considered to be the period from December through May. But the month of April was the wettest month over the southern latitude, when the number of rain events was 57 out of total number of 217 events during the year 2009. The number of rain events goes beyond 200 for rain rates up to 15 mm/hour, but this number drastically goes down to 75 for rain rates ranging between 15 and 25 mm/hour. The characteristic features of this type of rain are sudden onset, fairly short durations of heavy rain-fall, and relatively small sizes. The dominant rain-forming mechanism is considered to be, in this region, due to convective activity. However, sometimes the raining systems are associated with a cold front, and in this case, rain can stay for several hours and is with large cloud clusters.

In another communication, Norbury and White (1971) reported some information obtained over southeast England. They used three rain gauges (to be discussed in the next section) and examined distribution of durations of events where rain rates exceeded various thresholds continuously with a minimum duration of 30 seconds. Figure 8.3 shows the number of times per rain gauge-year those durations exceeded various periods for thresholds between 0.5 and 50 mm/hour. The data were fitted to a power law relation, following Ajayi and Ofoche (1984), as

$$N(R) = \alpha \, (\Delta t)^\beta \tag{8.1}$$

Here $N(R)$ is the number of occasions when a rain rate R was continuously exceeded for durations of Δt seconds. The constants were found to be

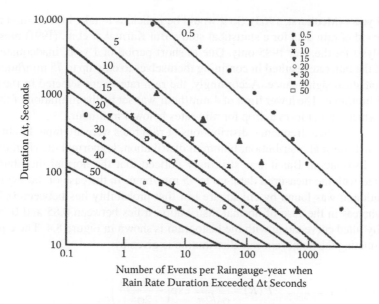

FIGURE 8.3   The number of times per rain gauge-year durations exceeded various periods.

$$\alpha = 4.2 \times 10^5 \, R^{-0.85}$$

$$\beta = -(1.37 + 0.02R) \tag{8.2}$$

for rain rates between 0.5 and 20 mm/hour.

But over a tropical location, according to Ajayi and Ofoche (1984), these are

$$\alpha = 3720 - 22.8R$$

$$\beta = 0.82 \, \exp(0.004R) \tag{8.3}$$

The reason for these types of relationships between tropical and temperate locations is yet to be explored very carefully. Now to extrapolate the study, we go back to make a comparison between Kolkata and the UK. There we see that the total number of rain events at Calcutta in the year 1986, with rain rates exceeding 25 mm/hour, is 85, while in the UK the total number is only 77 for rain rates exceeding 20 mm/hour, measured during a 4-year (1970–1973) period. This exhibits larger abundances of rain events over a tropical location like Kolkata than over a temperate location like the UK. The rain intensity data, as obtained in Kolkata (22°N), were analyzed by Karmakar et al. (1991) to find out the frequency distribution of rain intensity. It was noted that in this location a rain rate beyond 85 mm/hour is very scanty, which was also supported by the data obtained by the India meteorological department. But it should be emphasized that some of the components affecting rain attenuation may be a very slowly varying function that may take several years to repeat. For example, in New Jersey the return period of rain rate beyond 150 mm/hour was found to be about 5 years (Medhurst, 1965). So it would be judicious to take into account a very long period of rain data for a statistical study. But Karmakar et al. (1991) presented the analysis for the year 1985 only. Due to short period of 1 year, inadequate sampling of the rain rate resulted in confining themselves to only up to 75 mm/hour with some statistical significance. Accordingly, the rain rates were selected in the range of 4–75 mm/hour. The lower limit of 4 mm/hour was set by the limitation of the rain gauge, which cannot form a drop for rain rates below 4 mm/hour.

Several statistical frequency distributions of rain rates were attempted to fit properly, such as normal distribution, binomial distribution, lognormal distribution, and gamma distribution. But it was strikingly noticed that the normal distribution is well fitted with the measured data, up to 75 mm/hour. In the case of the lognormal distribution, it was found by chi-square test that probability lies between 0.30 and 0.50, whereas in the case of normal distribution it lies between 0.85 and 0.75. The normally fitted curve along with the histogram is shown in Figure 8.4. The equation approximating the point rain rate distribution is given by

$$P(R) = \frac{1}{\sigma\sqrt{2\pi}} \exp\left(\frac{-(R-\mu)}{2\sigma^2}\right) \tag{8.4}$$

Here μ and σ are the mean and standard deviation, respectively. In this case it was found that $\mu = 32.95$ mm/hour and $\sigma = 14.43$ mm/hour.

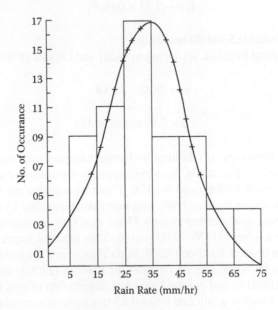

FIGURE 8.4 Normal distribution of rain intensity at Kolkata (22°N) up to 85 mm/hour during 1985 only. But beyond this 85 mm/hour the distribution appears to be lognormal. Due to the scarcity of data, a lognormal distribution was not attempted. However, during 1991–1992 the fitted distribution was found to be lognormal.

Also examined was to what extent rate may be expected to fall into each of various classes of frequency distribution. For example, to ascertain the proportion of rain rate that may be expected to go beyond 50 mm/hour, the proportion that may be expected between the values of mean $\mu = 32.95$ and $R = 50$ mm/hour was examined first, and then this proportion was subtracted from 0.50. At $\mu = 32.95$ mm/hour, the deviation from mean to the limit of the respective classes is $50 - 32.95 = 17.05$ mm/hour. Now if we look to a statistical table, it would be found that the expected area is 0.3810 between mean and 50 mm/hour, and therefore $0.5 - 0.38 = 0.12$, or about 12% is the probability of raining beyond 50 mm/hour, over Kolkata.

8.2.1 POINT RAIN RATES

Communication networking with the help of a millimeter-wave system needs characterization of a rain climatic region on the basis of rainfall intensity distribution. In this context, the data bank is still not free from scarcity. Variability is also expected from location to location within a certain region. However, the data pooled from different regions can provide a good statistical estimate. This in turn suggests a time-space ergodicity for the cumulative distribution of rainfall intensity.

In fact, cumulative distribution of rainfall intensity can be obtained with sufficient accuracy by means of a rapid-response rain gauge up to a very high rain rate. In recent times, the disdrometer is available, which may be called upon as a laser precipitation monitor. But one has to take care of the integration time of the gauge because it can affect the distribution pattern very much. A detailed description about this has been reported in much literature (Crane, 1996). However, the cumulative distribution of rainfall intensity over Kolkata (tropical location) is presented in Figure 8.5. Three different rain statistics are obtained from three different sources.

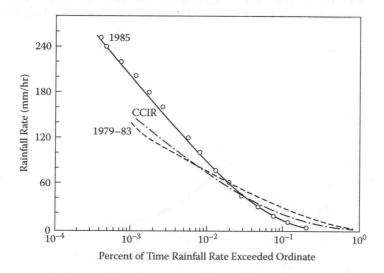

FIGURE 8.5 Cumulative distribution of rainfall intensity in Kolkata, obtained from rain gauges of different time constants.

The difference in cumulative distribution of three types of rain data is very small, although the integration for a rapid-response rain gauge is of the order of 15 seconds, and it is 1 minute for the comité consultatif international pour la radio (CCIR) study and 5 minutes for the tipping bucket rain gauge. Moreover, this measured similarity is well supported by the work of Fedi (1981). It is suggested there that for the 50–75 mm/hour rain rate, the integration has more or less no effect on cumulative distribution. But for good accuracy of the measured results, it is always suggested to use the rapid-response rain gauge with a smaller integration time of the order of 1 minute.

However, it is reported by Norbury and White (1971) that they have measured rainfall intensity over southeast England (temperate location) by exploiting three spaced rapid-response rain gauges, 200 m apart, in which rain water is collected in funnels of 150 cm^2 aperture and then formed into constant volume drops that are counted in 10-second intervals. The uncertainty in determining a uniform rain rate of 20 mm/hour was about 10% and decreased for higher rain rates. They collected data for a 3-year period from October 1982 to September 1985. Figure 8.6 shows the cumulative distribution of rainfall rates exceeded during this period, averaged over nine rain gauge-years. For comparison, the cumulative distributions for CCIR rain zones E and F (CCIR, 1986) are shown. For time percentage greater than 0.01%, the agreement with both zones is good, while at a smaller percentage of time, rain zone E appears more appropriate, although it is recognized that the differences are probably not significant due to inherent variability in meteorological processes, which can typically be of the order of 30–40%. Monthly variations were analyzed by averaging the monthly cumulative distributions from all three gauges and estimating the rainfall rate at different time percentages by quadratic interpolation for different time percentages. In most cases two such sets of the three values were interpolated, with the resulting interpolated values differing by 5% or less. In this way they have presented the mean monthly variation for time percentages between 1 and 0.001% (Figure 8.7). It shows that little monthly

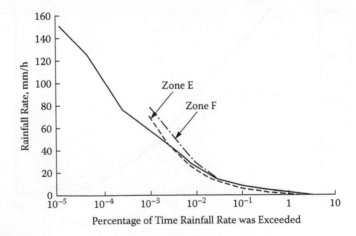

FIGURE 8.6 Cumulative distribution of rainfall rate from three rain gauges over a 3-year period. Also shown are the CCIR rain zones E and F.

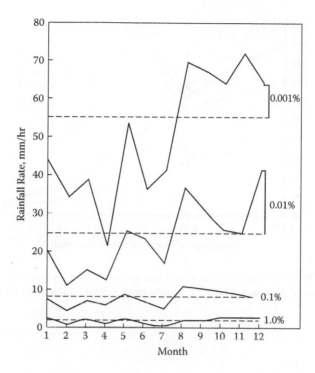

FIGURE 8.7   Monthly variations in rain gauges for different levels of probability in the UK.

variation in the lower rain rates occurs, and light rainfall is fairly evenly distributed throughout the year, while heavier rainfall tended to occur during the later
months of the year. Rain in these regions is mostly of stratiform structure, which
is generally light with relatively large rain cell diameters. But in the tropics, rain at
times is from convective rain cells with relatively small diameters, often resulting
in heavy downpours for short periods (Ramachandran and Kumar, 2005, 2007;
Mandeep and Allnut, 2007) reported the results obtained from experimental data
over the University Sains Malaysia (USM), Institute Technology Bandung (ITB),
Atendo de Manila University (AdMU), and University of South Pacific (USP).
The time exceedence curve in Figure 8.8 shows that as the rain rate increases,
the trend of the slope of the curve decreases gradually from a large negative, and
then the trend is reversed. Figure 8.9 shows the cumulative distribution of measured rain rate for the year 2011 over Fortaleza, Brazil. It is clearly noticed that
a sudden change in slope exists, which is again supported by Figure 8.10. The
change occurs at around 9–11 mm/hour. The existence of a "break point" in the
exceedence curve is denoted by an arrow. The presence of such break points in
the exceedence curve was reported by several authors (Pan and Allnutt, 2004;
Ramchandran and Kumar, 2007; Mandeep and Allnutt, 2007). The break points
here refer to the point at which the trend gets reversed (Kumar et al., 2006; Bryant
et al., 2001). When the rain structure is stratiform, rainfall is widespread, with
low rain rates (Mandeep and Allnutt, 2007). This usually occurs in the tropics;

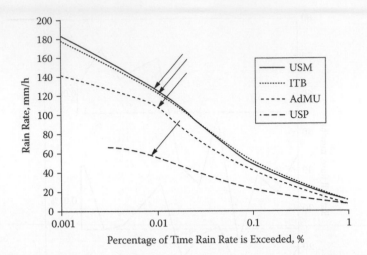

FIGURE 8.8 The exceedence curve shows that as the trend of the rain rate increases, the slope of the curve decreases gradually from a large negative value, and then the trend is reversed.

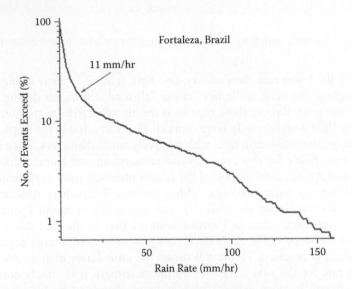

FIGURE 8.9 Cumulative distribution of rain rate of 1-month campaign data over Fortaleza.

when the cloud builds up, the water droplets are trapped in updrafts inside the cloud and are vertically transported. This enhances the coalescence of water particles, resulting in heavy rain (Schumacher and Houze, 2003). Thus, any change in the trend of the exceedence curve signifies that the rain structure changes from stratiform to convective rain.

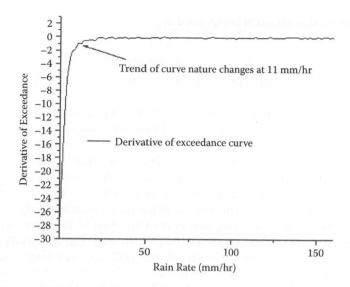

FIGURE 8.10 Derivative of cumulative distribution of rain rate (refer to Figure 8.5) of 1-month campaign data over Fortaleza.

8.3 RAINDROP SIZE DISTRIBUTION AT TROPICAL LOCATIONS

The measurement of raindrop size distribution (DSD) has continued interest in view of its role in determining the microwave properties of rain and the evolution of rain events, classifying types of rain and various climatic studies. The DSD data are still sparse in the tropical region, particularly in the tropical locations, in view of the complex climatic behavior of these regions.

The measurements of DSD (Maitra and Chakravarty, 2005) made at Kolkata (22°N) have been used to study the evolution of DSD during rain events and to indicate the different phases of the events. Also, an effort has been made to model the DSD data in terms of three-parameter distributions, namely, lognormal and gamma. The integral rainfall parameters (IRPs) are estimated from the measured and modeled DSDs, and they are intercompared to indicate the efficacy of the modeling of DSD. Also, the integral parameters obtained from the Marshall-Palmer (MP) distribution are compared to the present values to examine the adequacy of the available models to describe the DSD at Kolkata.

The DSD data are fitted to the three-parameter lognormal and gamma distributions. The gamma function is given by

$$N(D) = N_0 D^n \exp(-\Lambda D) \tag{8.5}$$

$$= N_T \Lambda^{n+1} D^n \exp(-\Lambda D)/\Gamma(n+1) \ m^{-3} mm^{-1} \tag{8.6}$$

Three distribution parameters, N_T (or N_0), Λ, and n, are considered to depend on the rainfall rate. $\Gamma(n+1)$ is the complete gamma function.

The lognormal distribution is expressed as

$$N(D) = \frac{N_T}{\sigma D \sqrt{2\pi}} \exp\left[-\frac{1}{2}\frac{(ln(D)-\mu)^2}{\sigma}\right] m^{-3} mm^{-1} \tag{8.7}$$

Here N_T, μ, and σ^2 are rain rate-dependent distribution parameters. The distribution parameters may be obtained by the method of moments (Ajayi and Olsen, 1985) and using the third, fourth, and sixth moments of the measured DSD.

Figure 8.11(a) shows the variation of DSD during a rain event at Kolkata on September 3, 2004, which shows the variation of the distribution of the fraction of total number of drops at different sizes. Figure 8.11(b) gives the variation of rain rate during the event. It is evident that the larger drops were relatively more abundant during the initial phase compared to the latter part of the event. As the rain event progressed, the fraction for smaller drops increased. The dominant drop sizes were in the range of 1 to 2 mm. The variation of DSD becomes more evident from Figure 8.12, in which the DSDs at two instances, 14:52:30 and 15:19:30 hours, are shown for identical rain rates around 38.5 mm/hour. At the earlier phase, the number of drops in the size range 3–4 mm was considerable, whereas smaller drops, in the

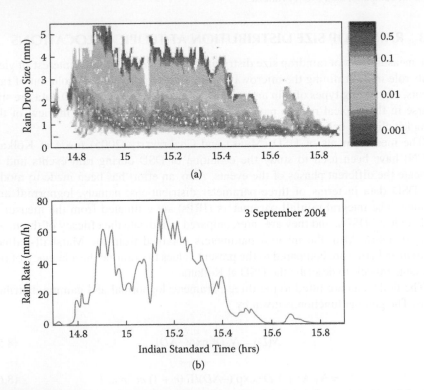

FIGURE 8.11 (See color insert.)   (a) Contour color map showing the variation of the distribution of the fraction of total number of drops at different sizes during a rain event on September 3, 2004. (b) Variation of rain rate during the same event.

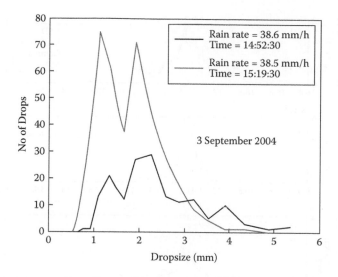

FIGURE 8.12 (See color insert.)   DSD at two instants of the rain event of September 3, 2005, for identical rain rates indicating a large variation of DSD at different phases of the event.

size range of 1–2 mm, significantly dominated in the later phase of the event. A dip in the rain rate around 15:06 hours (Figure 8.11) was caused mostly by the disappearance of the larger drops. Since the larger drops are associated with convective processes, and the dominance of smaller drops indicates the stratiform nature of rainfall, a continuous depiction of DSD during a rain event using disdrometer measurements can be useful in identifying the different precipitation processes during the event.

Figure 8.13 gives the variation of different integral rainfall parameters (IRPs), namely, liquid water content, radar reflectivity factor, and Ku-band specific attenuation, during the rain event of September 12, 2004. The curves depict the values of IRPs obtained from DSD measurements, fitted with lognormal distribution, gamma distribution, and Marshall-Palmer (MP) distribution. It is observed that the lognormal distribution provides the best match with the DSD-generated values. The gamma distribution slightly overestimates the DSD-generated values at low rain rates and underestimates at high rain rates. The MP distribution gives significantly higher estimates of IRP compared to these values. Since the radar reflectivity factor is determined by the sixth moment of DSD, the differences among various sets of data are most prominent for this parameter.

To understand the discrepancies shown by the models, a color map showing the variation of the difference between the measured and modeled DSD values during the rain event is given in Figure 8.14. A positive difference indicates a higher measured value and a negative one a lower measured value, compared to the modeled value. It is seen that both the lognormal and gamma distributions significantly underestimate the number of drops at small sizes (≤0.5 mm), which have less effect on IRP. The underestimation by the gamma model, however, extends to the higher sizes at higher rain rates, resulting in lower estimates of IRP by the gamma. The lognormal

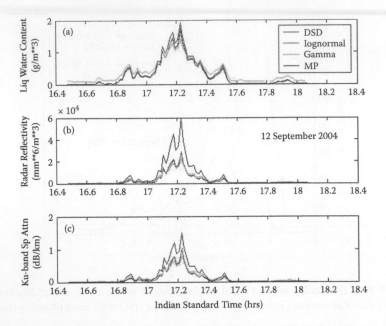

FIGURE 8.13 (See color insert.) Variation of different integral rainfall parameters (IRPs) obtained from DSD measurements, a fitted lognormal model, a fitted gamma model, and a Marshall- Palmer (MP) model.

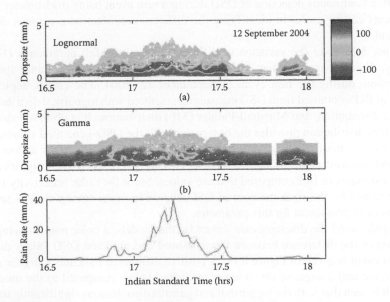

FIGURE 8.14 (See color insert.) Contour color map showing the difference between the measured DSD values and the modeled values with (a) lognormal and (b) gamma distributions during the rain event of September 12, 2004.

TABLE 8.1

Drop Size Distribution Parameter at Different Tropical Locations

Place	Latitude	N_T	μ	σ^2
Kolkata, India (Maitra, 2000)	22°N	$546\,R^{0.469}$	$\mu = -0.538 + 0.017 lnR$	$0.0689 + 0.076 lnR$
Guwahati, India (Timothi et al., 1995)	26°N	$180\,R^{0.39}$	$log\,\mu = 1.12 + 0.18 lnR$	$0.48 - 0.03R$
Dehradun, India (Verma and Jha, 1996)	30°N	$169\,R^{0.294}$	$\mu = -0.056 + 0.131 lnR$	$0.3 - 0.024 lnR$
Nigeria (Ajayi and Olsen, 1985)	7.46°N	$108 R^{0.363}$	$\mu = -0.195 + 0.199 lnR$	$0.137 - 0.013 lnR$
Fortaleza, Brazil	3°S	$142.31\,R^{0.6568}$	$\mu = 1.27 + 0.082 lnR$	$-1.477 + 0.955 lnR$

model provides a better fit, on the whole, over the size ranges that have the most significant contribution to the estimate of IRPs.

However, it is well understood that a comprehensive description of DSD in the tropical region is hindered by a lack of adequate experimental observations that will cover the prominent regional and seasonal variability of the average air motion and turbulence responsible for the development of precipitation and subsequent coalescence and breakup of raindrops (Maitra, 2000; Verma and Jha, 1996). Three models in terms of a lognormal function pertaining to three locations across the different places of India are published in the open literature.

The lognormal function representing number of drops N of diameter D per unit volume (refer to Equation 8.7) in several tropical locations is described in Table 8.1. This includes the recent observation in Fortaleza, Brazil. The fitted lognormal distribution at 11 mm/hour in Fortaleza, Brazil, is presented in Figure 8.15. The variation of $N(D)$ with drop diameter over the same place is shown in Figure 8.16. This shows that the maximum value of $N(D)$ goes up to 4500, having around a 0.5 mm raindrop diameter in Fortaleza, Brazil.

8.4 RAIN ABSORPTION MODEL

The critical role of the propagation impairment on microwave systems and lack of rain measurement data from tropical regions for modeling purposes have been concerns of many organizations, like the International Telecommunication Union (ITU), European Space Agency (ESA), and European cooperative program (COST). This has become necessary because of the peculiarity of the tropical regions, which are characterized by high-intensity rainfall, enhanced frequency of rain occurrence, and the increased presence of large raindrops when compared with temperate climate. Another very important effort toward gathering more information is through Tropical Rain Measurement Mission (TRMM), jointly developed by the United States and Japan, and the Global Precipitation

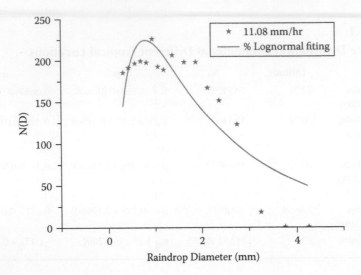

FIGURE 8.15 Lognormal distribution of raindrop size in Fortaleza, Brazil.

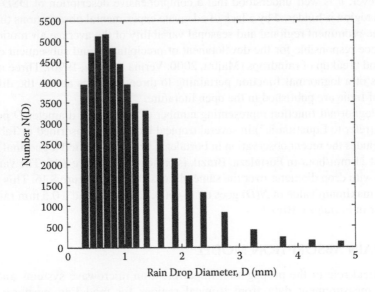

FIGURE 8.16 Histogram showing the value of $N(D)$ at different raindrop diameters over Fortaleza, Brazil.

Climatology Project (GPCP) of the World Climatic Research Programme (WCRP) (Ojo et al., 2008). It is important that the model is smooth, monotonic, and applicable to all time percentages, 0–100%. This is not quite as simple as it appears. Rain does not occur for much of the time, and when there is no rain, there is no rain attenuation. When rain does occur, the rain rate will not be uniform along the entire path. Further, at vanishingly small time percentages the rain rate cannot be allowed to become infinite but must saturate at some value. The time percentage below which rain does not occur at a specific location is pre-rain, an estimate of which is available from ITU-R P.837-5. We do not know what this maximum value of attenuation is, but clearly there must be some physical limit on the maximum rain rate. Both of these thresholds are functions of climate and link geometry. While it should be obvious that the attenuation will increase with the link length through rain, the probability of encountering rain along a path also rises with the path length; i.e., pre-rain attenuation decreases as the link length increases. The cumulative distribution function must, by definition, be monotonic. Although ITU-R P.837-5 provides the probability of rain, currently ITU-R P.618-9 assumes rain is only present for 5% of the time, and P.530-12 is only valid for less than 1% of the time.

It is worth briefly revisiting how these recommendations work out the rain attenuation. Both start by calculating the specific attenuation due to rain based on the relationship

$$A_s = kR^\alpha \quad \text{dB/km} \tag{8.8}$$

where R is the rainfall rate in mm/hour, and k and α are tabulated constants that depend on frequency and polarization. These constants are available in tabulated form and as a set of curve fits in Recommendation P.838-3. The tabulated constants were derived from theory, based on measured drop size and shape distributions, and account for both the scattering and absorption of the microwave wave front.

The overall attenuation is found by integrating the specific attenuation along the path

$$A_{total} = \int_0^d A(x).dx - \int_0^d kR(x)^\alpha .dx \tag{8.9}$$

The rain rate, and hence the attenuation, may vary significantly along longer paths, and for practical use a path average value is taken.

$$A_{total} = \int_0^d kR_{av}^\alpha .dx = d.kR_{av}^\alpha = d.A_{av} \tag{8.10}$$

The process of raindrop formation, growth, transformation, and decay occurs on a microphysical scale within a large cloud scale environment. Each process, such as condensation growth/evaporation or collision/coalescence, leaves a signature on the drop size distribution (DSD) of the rain event. Hence, analysis of the form of the DSD, its temporal and rain rate-dependent evolution at the surface and also aloft, is essential in understanding the process of rainfall formation. Recent DSD studies have focused on the differences between convective and stratiform rain, their differing characteristics, and the physics of their formation. Houghton (1968) pointed out that the primary precipitation growth process in convective precipitation is a collection of cloud water by precipitation particles in strong local updraft cores. As the parcels of air in the convective updrafts rise out of the boundary layer and reach the upper troposphere, they broaden and flatten as a result of decreasing vertical velocity of the parcel and, on reaching their level of neutral buoyancy, spread out and amalgamate to form a large horizontal area, which we identify as a stratiform region on the radar (Yuter and Houze, 1995). Under these conditions, the primary precipitation growth process is vapor deposition on ice particles formed earlier in the convective updrafts, but left aloft as the convective drafts weaken. It is a slow process, with particles always settling downward.

8.4.1 ATTENUATION MODEL PROPOSED BY THE UK

A new model including the path length adjustment factor was proposed by the UK. This takes the form

$$A = kRd.r(R,d)\qquad(8.11)$$

where $r(R,d)$ is the path reduction factor, a function of rainfall rate, R, and path length, d. The use of rain distribution rather than $R_{0.01}$ produces the path reduction factor as

$$r = \frac{1}{0.874+0.0255(R^{0.54}-1.7)d^{0.7}}\qquad(8.12)$$

Figure 8.17 shows the value of the function given in Equation 8.12, and the difference between the new method and the older one is shown in Figure 8.18. This model has been extensively tested against data and in general gives superior results.

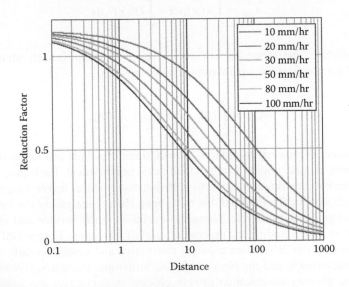

FIGURE 8.17 (See color insert.) UK 2003 path reduction factor.

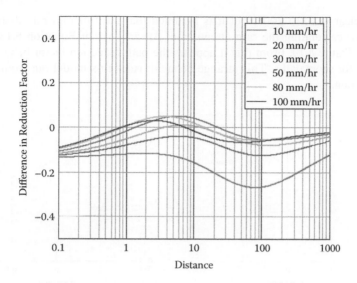

FIGURE 8.18 (See color insert.) Modified UK 2003 path reduction factor. It shows a better agreement than proposed in Figure 9.12.

8.4.2 Attenuation Model Proposed by the People's Republic of China

The People's Republic of China proposed a model in document 3J96 2005. While this was shown to be an improvement against the measurement database at 0.01% time, it does not cover the complete time percentage range. This model was of the form

$$r = \frac{1}{0.477d^{0.33}p^{0.073}\alpha f^{3.123} - 10.579(1 - g^{-0.024d})}$$ (8.13)

$$A_p = A_{0.01}\left[\frac{p}{0.01}\right]^{-\left[0.864 - 0.026ln\left(\frac{1+\mu}{p}\right) - 0.022ln(1+A_{0.01}) - 0.03ln(f) - 0.226(1+p)\right]}$$ (8.14)

The 3J53 2005 report contains a useful set of comparisons of this and alternative models, including that from the UK. China further developed a path length reduction factor model in 3J20 2008, based on an EXCELL rain model.

$$r = \frac{1}{0.19K^{-0.062}R_0^{0.31\alpha} - 1.7K^{0.63}\alpha R_0^{0.43\alpha}d^{0.48}(e^{-0.92R_0^{-0.027}}d^{0.16} - 0.5)}$$ (8.15)

where R_0 is the rainfall rate.

The model assumes there is only one rain cell along the path, with the constants in Equation 8.10 being found by a regression fit to the database. This fits well against the ITU-R measurement data bank using the ITU-R testing variable. It is not made entirely clear in 3J20 2008, but it is assumed that the full rain rate distribution is used for calculating the rain attenuation. An analysis of Equation 8.13 is presented in Figure 8.19. From this it appears the path reduction factor behaves very strangely for short paths. This strange behavior rules this model out from further consideration.

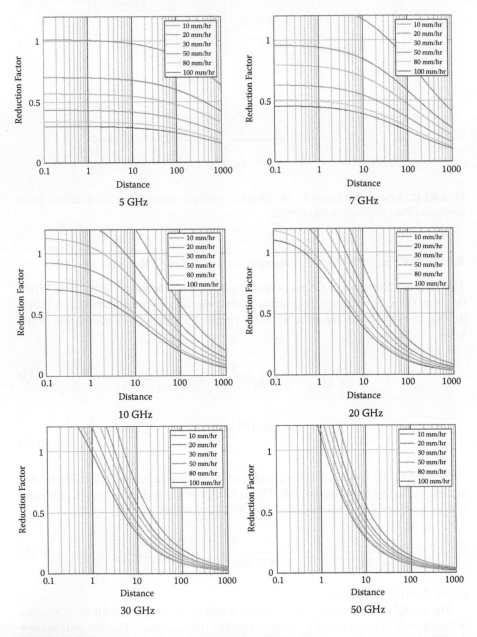

FIGURE 8.19 (See color insert.)   China path reduction factor.

8.4.3 Attenuation Model Proposed by Brazil

Brazil, in document 3J23 2008, provides a new model. It introduces the concept of an effective rain rate that applies to the whole path and depends on the path length, which is used in addition to a path reduction factor.

$$A_p = k\left[1.763R^{0.753+0.197/L_s\cos\theta}\right]^\alpha \frac{d}{1+\dfrac{d}{119R^{-0.244}}} \tag{8.16}$$

where θ is the elevation angle and the other terms have their usual definitions. This model has been tested against the ITU-R data bank and appears to give good results. A comparison was made with the UK model and the models proposed by Australia and China based on the standard test variable for rain attenuation given in Recommendation ITU-R P.311-12.

This model is similar in form to the UK 2003 model, assuming a link elevation of 0°, where Equation 8.16 simplifies to

$$A_p = k\left[1.763R^{0.768+0.197/d}\right]^\alpha \frac{d}{1+\dfrac{d}{119R^{-0.244}}} \tag{8.17}$$

It must be noted that the model from Brazil is not valid if d becomes small due to the $0.197/d$ term in the exponent of the effective rain rate.

This unfortunate behavior is demonstrated by Figure 8.20. While 3J23 acknowledges that the effective rain rate should be allowed to be larger, allowing it to approach infinity is not appropriate, and it is suggested that this model should not be used below 1 km.

The model from Brazil has a great advantage, as it is also applicable to satellite links, where it takes the form

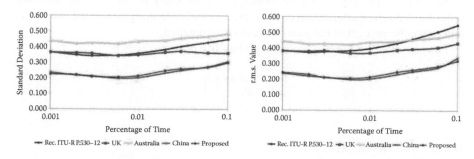

FIGURE 8.20 (See color insert.) Testing of Brazil model against terrestrial ITU-R database and other proposed models.

$$A_p = k[1.763R^{0.753+\frac{0.197}{L_s\cos\theta}}.\cos\theta + \frac{203.6}{L_s^{2.455}}R^{0.354+\frac{0.088}{L_s\cos\theta}}.\sin\theta]^\alpha \frac{L_s}{1+\frac{L_s\cos\theta}{119R^{-0.024}}} \quad (8.18)$$

Here the term L_s is the slant path length. The results are promising, though the improvement is small.

8.4.4 CRANE MODEL

This model is basically used to predict the attenuation due to rain on a terrestrial path. But care has to be taken about the endpoint rain climate. According to Crane, for path $x \le d \le 22.5$ km, the attenuation is

$$A = \alpha R_p^\beta \left\{ \left(\frac{e^{\mu\beta d}-1}{\mu\beta} \right) - \left(\frac{b^\beta e^{c\beta x}}{c\beta} \right) - \left(\frac{b^\beta e^{c\beta d}}{c\beta} \right) \right\} dB \quad (8.19)$$

But for $d < x$,

$$A = \alpha R_p^\beta \left(\frac{e^{\mu\beta d}-1}{\mu\beta} \right) dB \quad (8.20)$$

Here

$$\mu = ln(be^{cx})/x$$

$$b = 2.3R_p^{-0.17}$$

$$c = 0.026 - 0.03lnR_p$$

$$x = 3.8 - 0.6lnR_p$$

And d = path length in kilometers, R_p = rain rate in mm/hour, to be determined from the Crane table, and α, β = regression coefficients to be obtained from the table.

But, while applying the Crane model, it is customary to find out the rain rate required to produce attenuation equal to the thermal fade margin. Second, we are to determine the percent of the year this rain rate exceeded.

Now for $d > 22.5$ km, the modified probability of occurrence is

$$p_1 = p\left(\frac{22.5}{d} \right)$$

8.4.5 ITU-R MODEL

For latitudes $\ge 30^\circ$N or S, the attenuation exceeded for percentage of time p in the range 0.001 to 1.0% may be deduced from the following:

$$\frac{A_p}{A_{0.01}} = 0.12 p^{-(0.546+0.043\log_{10} p)} \tag{8.21}$$

This relation is found to be valid for determining the following factors: 0.12 for $p = 1\%$, 0.39 for $p = 0.1\%$, 1.0 for $p = 0.01\%$, and 2.14 for $p = 0.001\%$.

On the other hand, for latitudes $<30°$N or S, the attenuation exceeded for percentage of time p in the range 0.001 to 1.0% may be deduced from the following:

$$\frac{A_p}{A_{0.01}} = 0.07 p^{-(0.855+0.139\log_{10} p)} \tag{8.22}$$

This relation is found to be valid for determining the following factors: 0.07 for $p = 1\%$, 0.36 for $p = 0.1\%$, 1.0 for $p = 0.01\%$, and 1.44 for $p = 0.001\%$.

8.4.6 MODIFIED ITU-R MODEL APPLICABLE FOR THE TROPICS

It is discussed that in the tropics the rain structure frequently changes from stratiform to convective. Hence, there lies a chance to intercept more rain cells in the slant path. This needs to introduce a correction factor, C_f, to calculate attenuation for path elevation angles less than 60°, according to Ramchandran and Kumar (2007). The attenuation can be calculated as

$$A_B = kR_B^\alpha L_E C_f$$

Here A_B is the break point attenuation, and the correction factor C_f is given by

$$C_f = -0.002\theta^2 + 0.175\theta - 2.3$$

The percent exceedence at the break point (where the exceedence curve changes its slope) attenuation has to be calculated. Ramchandran and Kumar (2007) proposed that for tropical locations the value of $p_{0.01}$ is 0.021%. According to the ITU-R recommendation (P.618-8, 2003), the attenuation $A_{0.01}$ is calculated from $R_{0.01}$. But in this modified model $R_{0.01}$ is used to calculate $A_{0.021}$. In the modification proposed, p in the ITU-R model is replaced by $p = 0.011$, so that when $p = 0.021\%$, $A_{0.021} = A_B$, the break point attenuation. The modified expression is

$$A_p = A_B \left[\frac{p-0.011}{0.01} \right]^{-[0.655+\ln(p-0.011)-0.045\ln(A_B)-\beta(0.989-p)\sin\theta]} \text{dB, for } 0.021 \le p < 1 \tag{8.23}$$

As discussed earlier, in the tropics when the rain rate increases and approaches the break point, the corresponding rain structure changes. It has also been indicated by radar observation that the convective rain cells are surrounded by stratiform rain (Schumacher and Houze, 2003). So, in attenuation estimation the effective path length in the different regions should be considered. Ramchandran and Kumar (2007) show that the prediction model for attenuation should be related to the magnitude of attenuation at the break point, and beyond this the rain is assumed to be convective and droplets are spherical. However, the expression for attenuation beyond this break point can be viewed as a modification to the ITU-R model, as

$$A_p = A_B \left(\frac{p}{0.021} \right)^{-0.5\{0.655+0.033\ \ln p^2 - 0.03\ln p - 0.045\ln A_B - \beta(0.989-p)\sin\theta\}} \quad \text{dB, for } p \leq 0.021 \quad (8.24)$$

Equation 8.23 gives a gradual increase in attenuation with increasing rain rate, and Equation 8.24 shows the attenuation tending to saturation.

8.5　RAIN ATTENUATION STUDIES OVER A TROPICAL LATITUDE—A CASE STUDY

A number of prediction procedures have been developed for earth-space paths over the last decade that are applicable to temperate climate, but have been found to overestimate rain attenuations in tropical regions (Dissanayake et al., 1990). This overestimation of the predicted result is considered to be due to an incorrect estimate of the effective path length, essentially leading to an inaccurate estimation of the path attenuation. The developments of improved and more accurate rain attenuation models applicable to tropical climates thus require more experimental data from such tropical regions. It has been considered that the equivalent vertical path length through the rain is not equal to the physical rain height. The ITU-R has developed a model for the path length reduction coefficient for the horizontal projection of the path, L_G, with the vertical path equivalent to the height of $0°$ isotherms (Ajayi and Barbalisca, 1990). An empirical vertical reduction factor for earth-space paths has been proposed to derive an effective rain height from the height of the $0°$ isotherm during rainy conditions.

In order to study the attenuations in the microwave band, continuous measurements of sky noise temperatures at 22.234, 23.834, and 30 GHz were conducted using a ground-based radiometer during the year 2009 at Cauchuria, Paulista (22.57°S, 89°W), Brazil (Karmakar et al., 2011).

The experimental measurements were supported by the results obtained from a fast-response optical rain gauge co-located with the radiometer.

8.5.1　Theoretical Background

Rain attenuation is characterized by the nonuniformity of rainfall intensity, raindrop number density, size, shape, or orientation, and raindrop temperature, in addition to its intrinsic variability in time and space. Rain attenuation on an earth-space path may be expressed by the following relation:

$$A(s) = \int_0^\infty \lambda(s)\,ds \quad (8.25)$$

where ds represents the incremental distance from the ground along the earth-space paths under consideration, and $\lambda(s)$ is the specific attenuation (dB/km).

For practical application, however, the rain attenuation is approximated to a simple power law for a number of raindrop size distributions and temperatures in the form

$$A(s) = \int_0^{L_s} aR^b \, ds \tag{8.26}$$

where a and b are the coefficients that depend on frequency, temperature, and raindrop size distribution; R is the rain rate in mm/hour; and λ is substituted as aR^b, dB/km. The parameter L_G is the projection of the rainy earth-space path along the ground (Figure 8.21) and is given by

$$L_G = [(H_R - H_S)/\tan(\theta)] \tag{8.27}$$

where θ is the angle of elevation, H_R the rain height, and H_S the height from sea level. The parameter L_s is the slant path below the rain height and is given by

$$L_s = [(H_R - H_S)/\sin(\theta)] \tag{8.28}$$

The rain attenuation (dB) may be determined from the measured rain intensity data assuming that the rain is spatially uniform. However, in practice, rain is generally not uniform over the entire radio path. Therefore, the entire path may be divided into small incremental volumes δV, in which the rain rate is approximately uniform. According to Semplek and Turrin (1969); and Bodtman and Ruthoff, (1974), an approximate choice of δV was considered to be 1 m³ for meaningful measurement of very low rain intensity. An experimental study (Lin, 1975) indicates that heavy rain also has a finite structure of the order of m³. With this choice of δV, total path

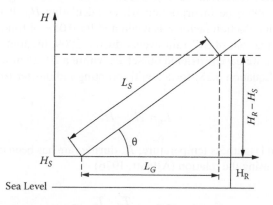

FIGURE 8.21   Schematic diagram of earth-space path.

TABLE 8.2

Values of Constants a and b

Frequency (GHz)	Value of a	Value of b
22.234	0.0766	1.1074
23.834	0.0906	1.1014
30	0.1581	1.0427

attenuation may also be obtained as the integral of the attenuation coefficient (dB/km) along the entire radio path.

The values of the coefficients a and b applicable to the chosen frequencies are taken from the ITU-R (ITU-R P.618-8, 1995) database, where the raindrop size was assumed to obey the lognormal distribution. These values are frequency and polarization dependent and are shown in Table 8.2.

In the simplified model, the rain intensity in the rain medium is considered not to vary along the path; i.e., the rain intensity is homogeneous along the vertical path up to a height H_R. This height is assumed to be the level, from which raindrops with a diameter larger than 0.1 mm fall, and may be described as the physical rain height. The rain attenuation in the zenith direction (z) is then given by

$$A(z) = (H_R - H_S)\, aR^b \text{ dB} \tag{8.29}$$

It may also be noted that the physical rain height is not easily measureable and is the simplest approximation identified with the $0°$ isotherm height.

The $0°$ isotherm height, i.e., the rain height during rainy condition for latitudes less than $36°$, is given by the relation (Fedi, 1981)

$$H_R = 3.0 + 0.028\Phi \text{ km} \tag{8.30}$$

where Φ is the latitude in degrees. For tropical latitudes, i.e., for $\Phi < 36°$, it has been proposed that a path reduction factor deduced in this regard using the ITU-R model (ITU-R P.618-8, 1995) be incorporated while calculating H_R. It is worthwhile to note that the path reduction factor was evaluated for 0.01% of time in a year. It was striking to note that out of 217 rain events during 2009, the highest rain rate was 107 mm/hour, but we have restricted ourselves within a 20–25 mm/hour rain rate.

Referring to Equation 8.28, for a zenith-pointing radiometer ($\theta = 90°$), we may write

$$L_s = H_R - H_S \tag{8.31}$$

The measured brightness temperature, T_a, during rain has been converted to total attenuation (dB) using the relation (Allnutt, 1976)

$$A = 10 \, \log_{10} \frac{T_m - T_c}{T_m - T_a} \text{ dB} \tag{8.32}$$

where T_m is the mean atmospheric temperature. For tropical latitudes like Brazil (22.57°S), the values of T_m will be higher than those in temperate latitudes, due to the higher temperatures and larger water vapor content (Sen et al., 1990), and is defined as (Wu, 1979)

$$T_m = \frac{\int T(z)\alpha(z)\exp\{\alpha(z)\,dz\}}{\int \alpha(z)\exp\left\{-\int\alpha(z)\,dz\right\}dz} \tag{8.33}$$

The value of T_m has been calculated with the help of a known vertical profile of atmospheric temperature and the corresponding vertical profile of attenuation coefficients. The value of T_c was considered to be 2.75 K (Ulaby et al., 1986). The attenuation coefficients are calculated by using the millimeter-wave propagation model (MPM) by Liebe (1989), where the input parameters were temperature, pressure, and humidity of the ambient atmosphere. These data were made available from the British Atmospheric Data Centre (BADC), over Brazil. To get the relation between the surface temperatures T_s, an attempt has been made, for the sake of simplicity, to correlate them with a linear relation, and subsequently it was found that $T_m = M + NT_s$ (Mitra et al., 2000), where M and N are the regression coefficients derivable for different frequencies. T_m is the mean atmospheric temperature dependent on frequency and is related to surface (ground) temperature. The values of M and N for 22.234, 23.834, and 30 GHz were calculated as $M = 270.05$, 270.03, and 270.00 K, respectively; $N = 0.778$, 0.791, and 0.816 K/°C. So, it is obvious that the values of T_m would be different for different frequencies. Hence, by observing the ground temperature and using the appropriate values of M and N, the mean atmospheric temperatures T_m were found. Now with the knowledge of T_m and T_c, and by using equation $T_m = M + NT_s$, the attenuation values were found where T_s is the surface temperature.

8.5.2   RAINFALL RATE MEASUREMENT

Figure 8.22 represents the histogram plot of rain intensity (mm/hour) measured by a fast-response disdrometer. It is interesting to note that the rain rate rarely goes beyond 100 mm/hour. There were a few events that occurred when the rain rate attained a maximum value of 107 mm/hour. A statistical analysis of the number of rain events over Cachoeira Paulista (22°S), Brazil, during the year 2009 reveals that the number of occurrences goes beyond 200 times for rain rates up to 15 mm/hour, but quite interestingly, it is also observed that the number of events suddenly falls to 75 for rain rates of 15–25 mm/hour. Hence, it suggests that at the place of experiment, the most abundant rain rate was up to 25 mm/hour.

8.5.3   BRIGHTNESS TEMPERATURE

The maximum brightness temperature observed by the radiometer was around 291 K for the water vapor frequency band (20–30 GHz). There were several events in which the sky noise temperatures exceeded 290 K. It is interesting to point out

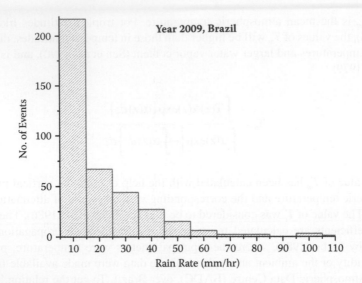

FIGURE 8.22   Rain rate distributions over Brazil.

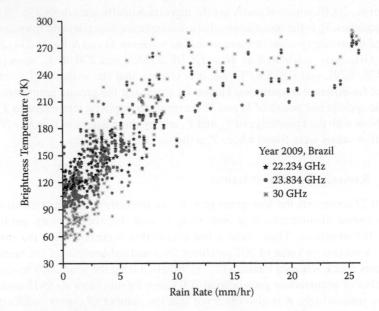

FIGURE 8.23 (See color insert.)   Variation of brightness temperature with rain intensity.

from Figure 8.23 that the brightness temperature sharply increases up to rain rate 10–12 mm/hour. But above this rain rate, the radiometer brightness temperature slowly increases and has a tendency of saturation at all frequencies. The relationships of the radiometer brightness temperature at different frequencies with rain rate appear to follow the power law equation, as is evident from Figure 8.23.

8.5.4 Attenuation

From the observed values of brightness temperature at three frequencies in the water vapor band, the vertical path attenuation was calculated. Figure 8.24 presents the time variation of measured and calculated path attenuations and the rainfall rate for a particular event on January 20, 2009. It is interesting to note that the variation pattern of path attenuation at all the frequencies followed a rain intensity (mm/hour) variation pattern measured by a fast-response disdrometer.

It should be mentioned that the disdrometer and radiometer were colocated at INPE, Brazil.

A slight departure is observed in the variation pattern of attenuation at 22.234 GHz (Figure 8.24(a)) during 21:35:00 to 23:43:25 hours (local time), when the rain rate was very low.

Below 5 mm/hour, the measured attenuation is found to be higher than the calculated attenuation. This might be due to the fact that the heated earth surface evaporates water vapor and subsequently is filling up the antenna beam by a larger amount of vapor when the raindrop falls on the surface. This becomes prominent only at 22.234 GHz, since the frequency of the radiometer lies just at the water vapor resonance line, although weak. But it is to be noted that the radiometer always maintained a threshold attenuation level. This shows the larger sensitivity of water vapor at 22.234 GHz, although this frequency is pressure broadened, and hence not suitable for accurate estimation of water vapor. This kind of anomaly is not recognized at 23.834 GHz (Figure 8.24(b)), which is found to be pressure independent. So it appears that rain attenuation measurements at 22.234 GHz are contaminated

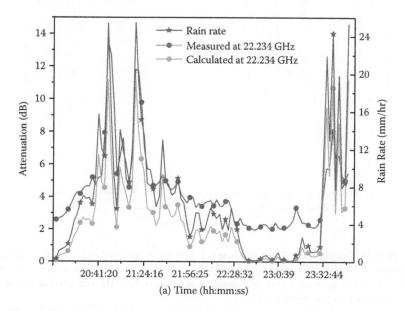

FIGURE 8.24 (See color insert.)   (a) Time series of measured and calculated attenuation at 22.234 GHz and corresponding rain rates over Brazil on January 20, 2009.

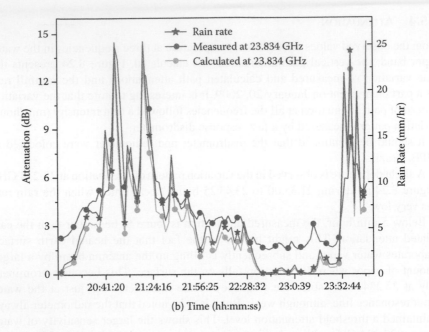

(b) Time (hh:mm:ss)

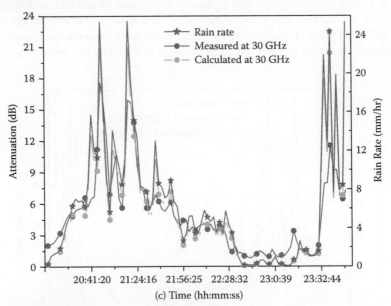

(c) Time (hh:mm:ss)

FIGURE 8.24 (*Continued*) (See color insert.) (b) Time series of measured and calculated attenuation at 23.834 GHz and corresponding rain rates over Brazil on January 20, 2009. (c) Time series of measured and calculated attenuation at 30 GHz and corresponding rain rates over Brazil on January 20, 2009.

by the unwanted presence of water vapor. This effect is minimized at the window frequency region (around 30 GHz) where attenuation due to water vapor is much less (Figure 8.24(c)).

Another interesting result is noticed here: at the peak rain rates the calculated attenuations at all the frequencies are much higher than those of the measured attenuations. It is very prominent at 30 GHz, as this is the highest frequency in this study.

Another attempt to work out regression analyses to a power law yields the following best-fit relations, at the said frequencies (Figure 8.25).

The equation was found to be in the form A (dB) = $H_R\, a(R - R_c)^b$.

$$A_{22.234}\ (\text{dB}) = 3.63 \times 0.367\ (R + 1.944)^{0.536}\quad (r^2 = 0.797)$$

$$A_{23.834}\ (\text{dB}) = 3.63 \times 0.332\ (R + 1.373)^{0.587}\quad (r^2 = 0.797)$$

$$A_{30}(\text{dB}) = 3.63 \times 0.353\ (R + 0.394)^{0.675}\quad (r^2 = 0.797)$$

The experimental results were compared with those obtained theoretically for different frequencies, using the values of a and b listed in Table 8.2. It was found that the calculated rain attenuations deviated significantly from the measured values, at higher rain rates. For this reason, best-fit curves have been drawn up to 20 mm/hour rain rates, and corresponding a and b coefficients matched very well up to this rain

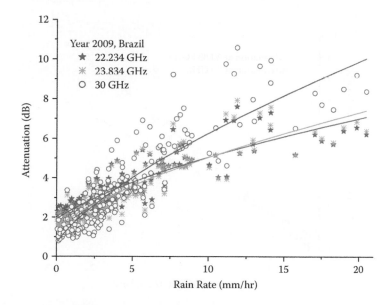

FIGURE 8.25 (See color insert.) Scatter and best-fit plot of rain attenuation at 22.234, 23.834, and 30 GHz and the corresponding rain rates.

rate. It should be noted that attenuation at all frequencies maintained a minimum threshold level. At 30 GHz it is nearly 0.685 dB, and at 22.234 GHz it is around 1.906 dB, even when the rain rate is zero. This is because the frequencies around 22.234 GHz are water vapor sensitive, but 30 GHz is not very sensitive to water vapor. So it is suggested that as we move on to the higher frequencies, from 22.234 up to 30 GHz, the water vapor sensitivity becomes less. Keeping this in mind, we were prompted to see the variation pattern of rain attenuation at 23.834 and 30 GHz with respect to 22.234 GHz (Figure 8.26) analyses of attenuation. Taking 22.234 GHz as the reference frequency, we get

$$A_{23.834} \text{ (dB)} = 1.133 \ (A_{22.234} - 0.6204)^{1.003} \quad (r^2 = 0.999)$$

$$A_{30} \text{ (dB)} = 1.823 \ (A_{22.234} - 1.528)^{0.982} \quad (r^2 = 0.997)$$

So using these equations, one can have the idea of getting an approximate value of rain attenuation at the said two frequencies by measuring rain attenuation at 22.234 GHz, at the corresponding rain rate. An attempt has also been made to present the variation of rain attenuation with frequency taking rain as a parameter (Figure 8.27). Here we have restricted ourselves within the rain rate 5–25 mm/hour. It is surprisingly noticed that rain attenuation increases monotonically with rain rate, with an exception at 23.834 GHz. During our study it is also observed that at 23.834 GHz the attenuation is always minimum, irrespective of any rain rate within the water vapor band. In this connection it should be mentioned that this 23.834 GHz is such

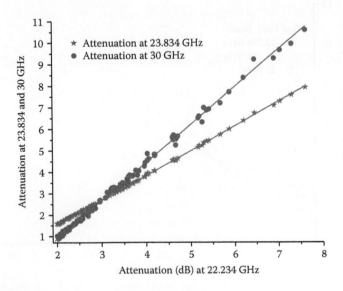

FIGURE 8.26 Scatter plot and best-fit result of attenuation at 22.834 and 30 GHz against 22.234 GHz during the year 2009.

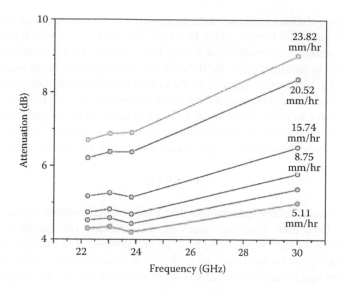

FIGURE 8.27 Variation of rain attenuation at different frequencies, taking rain as a parameter.

a frequency that is found to be pressure independent and considered to be a good choice of radiometric measurement in the water vapor band.

8.5.5 RAIN HEIGHT

The nonuniform horizontal rain structure is accounted for in the use of a rain rate reduction factor to convert the physical path length to an effective path length. The simple vertical structure assumes that rainfall is uniform from the ground to rain height. The physical rain height is the level up to which the water drops with diameters larger than 0.1 mm are present. However, the effective rain height may be obtained by analyzing the measured attenuation and point rainfall intensity data. Any nonuniformity of the vertical profile of rain is, in fact, integrated with time provided that all the water falling inside an ideal column ultimately reaches the ground. But it may so happen that a few raindrops remain aloft, and then the radio wave propagation may not be affected. Moreover, wind may also drive away the floating raindrops from the radio path. But in practice, the vertical nonuniformity is very unlikely to occur, and it causes the effective rain height to seem higher than the physical rain height. The situation becomes more complex when horizontal nonuniformity occurs and is relevant in considering the global rain attenuation effect. At first approximation, the vertical nonuniformity may be disregarded in comparison to horizontal nonuniformity, except for high elevation angle ($\varphi > 60°$). This is the reason for which the rain height may be used instead of rain thickness for both physical and effective measurements of rain attenuation. The physical rain height is not easily measurable, and the closest approximation for rain height is to consider the $0°$ isotherm, which is readily available from radiosonde data. But during rain, there lies a big difference between the two, which in turn depends on the types of rain (Schumacher and Houze, 2003). In warm rain, the physical rain height is lower than the 0° isotherm. In thunderstorm rain, it is normally present well above the 0° isotherm, and in stratiform rain, the 0° isotherm and the physical rain height become coincident. This happens especially in the cold season when the falling ice crystals melt below the 0° isotherm height (Ajayi and Barbaliscia, 1990). However, we have taken the liberty to calculate or to get a firsthand idea of rain height from the measured rainfall intensity data corresponding to rain attenuation data at 22.234, 23.834, and 30 GHz. From Equation 8.31, the attenuation can be expressed in the form

$$A = aR^b H_R \qquad (8.34)$$

where H_R is the effective rain height. The experimental measurements of attenuations have been used to derive values for this effective rain height over Brazil. Figure 8.28 shows the obtained results at all frequencies. The average rain heights at different frequencies are shown in Table 8.3. It is interesting to point out that the effective rain height measured at 23.834 matches well with the physical rain height, which is 3.62 km.

The studies presented here give an idea of rain attenuation measurement by radiometer applicable to a tropical climate. From the measured result it was found that extra attenuation caused by evaporation of water vapor from the heated earth surface at the time of the start of rain, particularly at low rain rates, below about 5 mm hour^{-1}, dominated the measured attenuation, which is an extra error in the measurement. The several experimental results over the different parts of the world, including temperate and tropical regions, reveal that the 0°C isotherm varies with several factors. In this context, Ajayi and Barbaliscia (1990) made a comprehensive study. There, the quantities h_{FM}, h_{FY}, h_{FS}, and h_{FR}, the mean values of the 0°C isotherm height in an average month, year, summer, and half-year, respectively, and the mean values for rainy conditions for various rain thresholds were taken into consideration. In the northern hemisphere the summer half-year includes the months from May to October, but in the southern hemisphere it is from November to April. In the northern hemisphere the results were obtained from 3.4 to 46°N, and those in the southern hemisphere varied from 6.88 to 45.47°S.

TABLE 8.3
Measured Rain Height by Radiometer at Different Frequencies

Frequency (GHz)	Rain Height (km)
22.234	4.572
23.834	3.80
30	2.94
Average rain height = 3.77 (km)	

It is observed also that there lies a negligible difference between noon and midnight values of h_{FY} and h_{FS}. The diurnal variations over the temperate and tropical locations were found to be insignificant in comparison to monthly or seasonal variation. This suggests that 1-year data are adequate for studying the year-to-year variability of a 0°C isotherm height over a particular place of choice. Similarly, in a tropical location like Minna, the monthly variation over a year is less than 5%. These confirm the negligible dependence of rain rate on the 0°C isotherm height. But the determination of h_{FR} during rainy conditions is difficult. Here it has been assumed that the significant rain occurs in the summer half-year. It has been observed that for the temperate location, when summer rains are considered alone, the 0°C isotherm height appears to be almost independent of rain intensity, at least up to 15 mm/hour (Ajayi and Barbalisca, 1990). But beyond this, there might be little dependence on rain intensity. So it is suggested to perform the rain height determination experiment within the limit of 10–15 mm/hour (Karmakar, 2012).

The mean value for the effective rain height measured at three frequencies, 22.234, 23.834, and 30 GHz, was deduced from 4.572, 3.80, and 2.94 km, respectively. Only the 30 GHz measurement is underestimated from the predicted rain height at 3.62 km. From this phenomenon it can be concluded that while measuring the rain-induced attenuations, one has to eliminate completely the water vapor contribution, and at the same time the drop size distribution has to be properly chosen. However, more data are required on rain attenuations in tropical locations in order to give better insight into the dependence of rain height on the rain type before any firm conclusion can be drawn.

8.5.6 Effect of Scattering by Rain Cells

It is presumably considered that rain is present over a microwave aerial, situated at ground level. Energy can reach the aerial directly from a source within its beam, and through single and multiple scattering by rain drops. There are four principal sources of such radiation: the rain, the ground, the clear atmosphere and the Sun. Here, we put emphasis on rain only (Karmakar and Halder, 2013). The aerial is assumed to be lossless with a very narrow beam and no side lobes. At frequencies in the millimetre and microwave bands, attenuation and emission from clouds are small compared with those from rain (Zavody 1974).

8.5.6.1 Properties of Rain

The rain effective temperature T_R is taken to be homogeneous and to extend everywhere from ground level to a height h. The drops are assumed to be randomly distributed in space and spherical, so that all the energy emitted and re-scattered is randomly polarized. It is also assumed that the log normal drop-size distribution holds at all heights at Fortaleza, Brazil (Karmakar and Halder 2012) and that the drops fall with their terminal velocity with respect to the ground. Radar evidence (Joss *et al.* 1968) indicates that the drop size distribution normally does not vary appreciably with height.

The contribution T_{AR} to the aerial temperature by emission from the rain is given by

$$T_{AR}(\tau) = T_{AR^0}(\tau) + T_{AR^1}(\tau) + \ldots = \sum_{n=0}^{\infty} T_{AR^n}(\tau)$$

$T_{AR^0}(\tau)$ is due to the energy emitted by the rain within the beam of the aerial, and the energy that has been emitted by the rain and scattered n times before being incident on the aerial gives rise to T_{AR^n}.

Scattering and absorption coefficients of rain can be calculated using the Mie theory. Mie Theory (Mie 1908), allows for exact modeling of wave propagation, absorption, scattering and hence attenuation characteristics of spheres and of cloud particles, provided that the dielectric and magnetic properties of the particles are known. Water droplets of rain and clouds are nearly spherical, homogeneous, and thus Mie Theory is quite appropriate for applications in atmospheric physics (analysis of radiometer data). Mie Theory has also been widely used to complement the results of Rayleigh scattering which are much simpler formulation, but applicable for small size parameters ($x = ak \ll 1$, $a =$ drop radius, $k =$ wave number) only (Deirmendjian 1969).

Mie Theory is needed in microwave radiometry of rain at all microwave frequencies. For scattering and backscattering, on the other hand, the Rayleigh Approximation is valid (deviations < 25%) up to D = 6 mm, thus covering the full range of rain drops. It is to be noted that at Fortaleza It has been found the maximum drop diameter is of 5 mm with a maximum population of 1mm at rain rate 10 mm/hr (Karmakar and Halder 2012). Since radiometer data are usually needed with an accuracy of about 1 dB (26%), the Rayleigh Approximation is sufficient for radiometer observations at GHz range (Mätzler 2002).

8.6.6.2 Radio Emission by Rain

Emission by rain within beam of aerial-scattering of zero order

We ignore any energy that has been scattered into the aerial. The power per unit bandwidth emitted by volume ΔV (refer to Figure 8.28) is given by

$$\Delta P = J \, \Delta V \tag{8.35}$$

Where J is the emission coefficient.

It follows from Kirchhoff and Rayleigh-Jeans Laws (Kraus 1966) that

$$J = \frac{8\pi}{\lambda^2} K T_R Q_a \tag{8.36}$$

Where K is Boltzmann's constant, λ is the wavelength and Q_a is total absorption coefficient of rain.

Absorption and scattering between ΔV and point B give rise to attenuation, and the power flux density per unit bandwidth is

$$\Delta W = \Delta P \Big/ \left[4\pi \{(Z' - Z)/\cos\theta\}^2 \right]^{\exp[Q_e(Z'-Z)/\cos\theta]} \tag{8.37}$$

Where Q_e is total extinction coefficient.

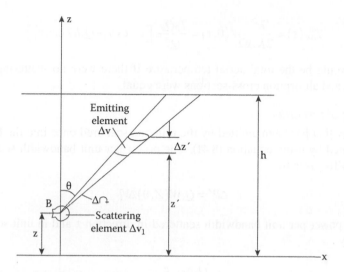

FIGURE 8.28 Co-ordinate system for calculating direct emission from rain.

Now we know

$$\Delta V = \left\{ (Z' - Z)/\cos\theta \right\}^2 \Delta Z' \Delta\Omega/\cos 0$$

Where Ω is solid angle,
Hence,

$$\Delta W = J\Delta Z' \, \Delta\Omega/\cos\theta \left[\left\{ -Q_e \left((Z' - Z)/\cos\theta \right) \right\} \right] \tag{8.38}$$

$$W(Z,\theta) = \int_z^h dW = \frac{2KT_R Q_a \Delta\,\Omega}{Q_e \lambda^2} \left[1 - \exp\{-Q_e (h - Z)/\cos\theta\} \right] \quad for\ \theta \leq \pi/2 \tag{8.39}$$

And

$$W(Z,\theta) \int_z^h dW = \frac{2KT_R Q_a \Delta\,\Omega}{Q_e \lambda^2} \left[1 - \exp\{-Q_e Z/\cos\theta\} \right] \quad for\ \theta \geq \pi/2$$

As T_R, and $Q_a \, Q_e$ are not functions of position, equation (8.39) can be integrated to get the total power flux density per unit bandwidth from direction θ at height. At ground level, for aerial elevation $\dfrac{\pi}{2} - \tau$,

$$W = (0,\tau) = \frac{2KT_R Q_a \Delta\,\Omega}{Q_a \lambda^2} \left[1 - \exp\{-Q_e h/\cos\tau\} \right] \tag{8.39}$$

The aerial temperature corresponding to the flux density above is given by

$$T_{AR}^0(\tau) = \frac{\lambda^2}{2K\Delta\Omega} W(0,\tau) = \frac{T_R Q_a}{Q_e} \left[1 - \exp\{-Q_e h / \cos\tau\}\right] \qquad (8.40)$$

and this would be the total aerial temperature if there were no scattering; i.e. the extinction and absorption cross-sections were equal.

Scattering of first order

The energy that has been emitted by the rain and scattered once into the beam can be calculated by using equation (8.41). The power per unit bandwidth scattered by volume $\Delta V1$ is given by

$$\Delta P' = Q_s W(Z,\theta) \Delta V_1 \qquad (8.41)$$

So the power per unit bandwidth scattered in direction τ and in unit solid angle is given by

$$P_R^1(Z,\tau) = \frac{Q_s \Delta V_1}{4\pi} \int S(\theta,\tau) W(Z,\tau) \qquad (8.42)$$

Where the integration is with respect to entire solid angle, S, θ, is the normalized scattering.

Substitution of W_Z, from equation (8.39) and integration with respect to θ gives the power per unit bandwidth propagating in direction τ in unit solid angle which is due to scattering by unit volume at height z:

$$P_R^1 = \frac{KT_R Q_a Q_s}{\lambda^2 Q_s} \int_0^{\pi/2} S(\theta,\tau)\sin\theta\left(1 - \exp\{-Q_s(h-Z)/\cos\theta\}\right)d\theta$$

$$+ \int_{\pi/2}^{\pi} S(\theta,\tau)\sin\theta\left\{1 - \exp\left(\frac{Q_e Z}{\cos\theta}\right)\right\}d\theta \qquad (8.43)$$

Where Q_s is the total scattering coefficient.

To obtain the contribution to aerial temperature by first order scattering and the full depth of the rain, a procedure similar to the previous section can be followed, but the "emission coefficient" is now given by

$$J' = 4\pi P_R^1(Z,\tau) \qquad (8.44)$$

and the aerial temperature is given by

$$T_{AR}' = \frac{\lambda^2}{2K\cos\tau} \int_0^h \exp\left(-\frac{Q_e Z}{\cos\tau}\right) P_R^1(Z\cdot\tau)dZ \qquad (8.45)$$

Approximated Scattering for second order

Scattering by particles similar to or larger than the wavelength is typically treated by the Mie theory, the discrete dipole approximation and other computational techniques. Rayleigh scattering applies to particles that are small with respect to wavelength, and that are optically "soft"(i.e. with a refractive index close to 1).

Rain drops are typically two orders of magnitude larger in diameter than cloud droplets. Hence, whereas for clouds the Rayleigh approximation is valid up to about 50 GHz (most cloud types) or even up to 300 GHz (fair-weather clouds), for rain the validity of rain approximation is limited to rain fall rates of less than 10 mm per hour in the centimeter wavelength range, and to very low rainfall rates at frequencies above 30 GHz.

So for the analysis of scattering caused due to rain drops, from the radiometric data, Rayleigh scattering is preferred and by modifying the first order scattering equation with proper limits we get the Rayleigh scattering co-efficient for a single particle which is given by the following expression (http://augerlal.lal.in2p3.fr/pmwiki/uploads/Bucholtz.pdf)

$$\sigma_s = \frac{2\pi d^2}{3\lambda^4}\left[\left(n^2-1\right)/\left(n^2+2\right)\right]^2 \tag{8.46}$$

Where σ_s = Rayleigh scattering co-efficient for a single particle in N_p/m
d = diameter of the particle; λ = wavelength; n = refractive index of the particle

Now the unit of scattering coefficient is N_p/m and hence in order to convert the unit to dB/Km, σ_s is to be multiplied by a factor 4.34 × 103 (Karmakar 2011). In order to obtain the scattering, due to rain drops, in dB the scattering coefficient values are to be multiplied by the rain height which is approximately 3.6 Km (Karmakar et al. 1991). Thus the total Rayleigh scattering (σ_T) for the rain particles with diameter d is the total number of rain particles with particular diameter d (N_d) times σ_s.

The total Rayleigh scattering for particles with particular diameter d is given by the expression

$$\sigma_T = \sigma_s \times 4.34 \times 10^3 \times 3.6 \times N_d \tag{8.47}$$

σ_T = Total Rayleigh scattering for particles with particular diameter d in dB
N_d = Number of particles with diameter d

8.6 NUMERICAL ANALYSIS

As a part of the objective of the CHUVA project, the rainfall intensity data were classified into several categories namely 0–15mm/hr, 15–30 mm/hr and so on. According to Karmakar and Halder (2012) the rain rate at Fortaleza, Brazil, shows that the maximum numbers of rain fall are at around 10 mm/hr. Hence the analyses were focused on the data of DSD at 10 mm/hr. It also shows that on April 9, 2011, during a 1-hour period the rain rate was mostly stable at around 10 mm/hr over Fortaleza. This was observed also that this eventually occurs many a times around 10 mm/hr over Fortaleza. The attenuation (dB) were calculated using the following equation (Allnutt, 1976)

$$A = 10 \log_{10} \frac{T_m - T_c}{T_m - T_b}$$

Here T_m is the frequency dependent mean atmospheric temperature (Maitra et al., 2002) and in fact, there is no appreciable variation in T_m in 20–30 GHz band. Hence $T_m = 297$K is considered. T_c is the cosmic background temperature = 2.7K and T_b is the radiometric brightness temperature.

Figure 8.29 shows the variation of attenuation with frequency (in the frequency range of 22 GHz to 30 GHz) at two different rain rates 10 mm/hr and 20 mm/hr. It is also clear that attenuation increases with increasing frequency in the frequency range 22 GHz–30 GHz and also the attenuation value increases with increasing rain rate with a slight departure at around 26 GHz exhibiting minimum attenuation in 20–30 GHz band.

We know the total Rayleigh scattering (σ_T) for the rain particles with diameter d is the total number of rain particles with particular diameter d (N_d) times σ_s (refer equation 8.47). The following table gives the list of N_d for respective values of d. The mentioned values of N_d with different diameters d are obtained by applying Log-normal fittings to the plot of number of particles per unit volume per unit diameter $N(D)$ versus diameter d (Karmakar and Halder 2012). Thus total scattering for different drop diameter values are obtained using equation (8.47) and variation of scattering in the frequency range 22 GHz–30 GHz are studied. Figure 8.30 shows the scattering versus frequency plots in the water vapor band with diameter as a parameter. Figure 8.30 prompts us to conclude that scattering increases with increasing frequency and also in reference to the above figure we can also say that scattering also increases with increasing drop diameter. Figure 8.31, which is a plot

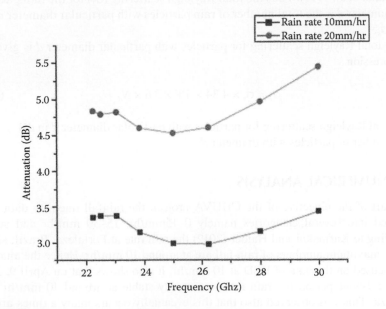

FIGURE 8.29 Attenuation versus frequency at rain rates 10 mm/hr and 20 mm/hr over Fortaleza, Brazil, on April 9, 2011.

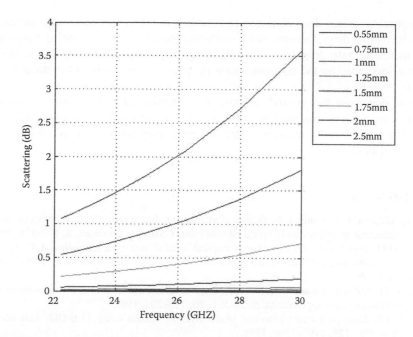

FIGURE 8.30 Scattering versus frequency curve for different drop diameters over Fortaleza, Brazil.

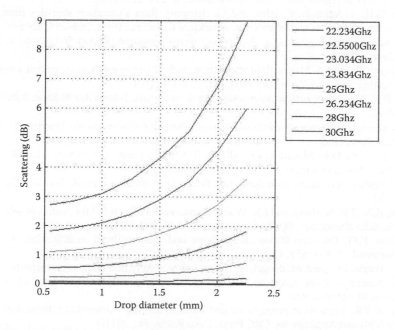

FIGURE 8.31 Scattering versus drop diameter for different frequencies over Fortaleza, Brazil.

of scattering versus drop diameter, re-establish this fact that scattering increases with increasing drop diameter. Since scattering increases with increasing frequency and increasing rain drop diameter in the frequency range 22GHz-30GHz so this is a great loss faced by the communication people while working in the water vapor band.

Emission from the ground can also be a significant factor while measuring attenuation. So it is suggested two radiometers, one situated over the sea and other over the ground could measure different aerial temperatures for rain with same characteristics.

REFERENCES

Ajayi, G.O., and F. Barbalisca. Prediction of attenuation due to rain: Characteristics of the 0° isotherm in temperate and tropical climates. *Int. J. Satellite Commun.*, 8, 187–196, 1990.

Ajayi, G.O., and E.B.C. Ofoche. Some tropical rainfall rate characteristics at Ile-Ife for microwave and millimetre wave applications. *J. Climate Appl. Meteorol.*, 23, 562–567, 1984.

Ajayi, G.O., and R.L. Olsen. Modelling of a tropical rain size distribution for microwave and millimetre wave applications. *Radio Sci.*, 20, 193–202, 1985.

Allnutt, J.E. Slant path attenuation and space diversity results using 11.6 GHz radiometer. *Proc. IEE*, 123, 1197–1200, 1976.

Bodtman, W.F., and C.L. Ruthoff. Rain attenuation on short radio paths: Theory, experiment and design. *Bell Syst. Tech. J.*, 55, 1329, 1974.

Bowthorpe, B.J., F.B. Andrews, C.J. Kikkert, and P.L. Arlett. Elevation angle dependence in the tropical region. *Int. J. Satellite Commun.*, 8, 211–221, 1990.

Bryant, G.H., I. Adimula, C. Riva, and G. Brussard. Rain attenuation statistics from rain column diameters and heights. *Int. J. Satellite Commun.*, 19(3), 263–283, 2001.

CCIR. *Radiometeorological data: Report and recommendation.* Report 563-3, vol. V. ITU, Geneva, 1986.

Chakravarty, K., and A. Maitra. Rain attenuation studies over an earth-space path at a tropical location. *J. Atmos. Terrestrial Phys.*, 72, 135–138, 2010.

Crane, R.K. *Electromagnetic wave propagation through rain.* John Wiley & Sons, New York, 1996.

Deirmendjian, D. *Electromagnetic Scattering on Spherical Polydispersions,* American Elsevier, New York, 1969.

Dissanayake, A., D.K. Mccarthy, J.E. Allnutt, R. Shepherd, and A. Restburg. 11.6 GHz rain attenuation measurements in Peru. *Int. J. Satellite Commun.*, 8, 229, 1990.

Fedi, F. Prediction of attenuation due to rain fall on terrestrial links. *Radio Sci.*, 16, 731–743, 1981.

Harden, B.N., J.R. Norbury, and J.K. White. Measurement of rain fall for studies of millimetric radio attenuation. *Optics Acoustics*, 1(6), 197–202, 1977.

Houghton, H.G. On precipitation mechanisms and their artificial modification. *J. Appl. Meteorol.*, 7, 851–859, 1968.

Joss, J., Thams, J.C., and Waldvogel, A. The accuracy of daily rainfall measurements by radar, Proceedings of the American Meteorological Society 13th radar meteorology conference, Montreal, 1968.

Karmakar, P.K. Microwave propagation and Remote Sensing: Atmospheric Influences with Models and Applications, CRC Press, Boca Raton, FL, 2011.

Karmakar, P.K and Tuhina Halder, Effect of scattering by rain on radiometer measurements, Novus Natural Science Research, Vol. 2, No.1. 11–20, 2013.

Karmakar, P. K., and Tuhina Halder. Rain drop size distribution at southern latitude, *Novus Natural Science Research*, Vol. 1, No. 4, 9–18, 2012.

Karmakar, P.K., R. Bera, G. Tarafdar, A. Maitra, and A.K. Sen. Millimeter wave attenuation in relation to rain rate distribution over a tropical station. *Int. J. Infrared Millimeter Waves*, 12(11), 1333–1348, 1991.

Karmakar, P.K., Chattopadhyay, S., and Sen, A.K. Millimeter wave attenuation in relation to rain rate distribution over a tropical station, International Journal of Infrared and Millimeter Waves (USA), 1991.

Karmakar, P.K., M. Maiti, C.F. Angelis, and L.A.T. Machado. Rain attenuation studies in the microwave band over a southern latitude. *Pac. J. Sci. Technol.*, 12(2), 196–205, 2011.

Kraus, J.D. Radio astronomy, McGraw-Hill, 1966.

Kumar, V., R.C. Deo, and V. Ramachandran. Total rain accumulation and rain analysis for small tropical Pacific islands: A case study of Suva, Fiji. *Atmos. Meteorol. Soc.*, 7, 53–58, 2006.

Kummerow, C., W. Barnes, T. Kozu, J. Shiue and J. Simpson, 1998, The tropical rainfall measuring mission (TRMM) sensor package, *J. Atmos. Oceanic Technol.*, 1998.

Liebe, H.J. MPM—An atmospheric millimeter wave propagation model. *Int. J. Infrared Millimeter Wave*, 10, 631–650, 1989.

Lin, S.H. A method of calculating rain attenuation distribution on microwave paths. *Bell Syst. Tech. J.*, 54, 1051, 1975.

Mandeep, J.S., and J.E. Allnutt. Rain attenuation prediction at Ku-band in Southeast Asia countries. *Progr. Electromagnetic Res.*, 76, 65–74, 2007.

Mätzler C., MATLAB Functions for Mie Scattering and Absorption, IAP Res. Rep. No. 02–08, 2002.

Medhurst, R.G. Rainfall attenuation of centimetre waves: Comparison of theory and measurement. *IEEE Trans. Antennas Propagation*, AP-13, 550–564, 1965.

Maitra, A. Three parameter raindrop size distribution modeling at a tropical location. *Electron. Lett.*, 36, 906–907, 2000.

Maitra, A., and K. Chakravarty. Raindrop size distribution measurements and associated rain parameters at a tropical location in the Indian region. Presented in URSI General Assembly, 2005.

Mie G., Beiträgezur Optiktrüber Medien, speziellkolloidaler Metallösungen, *Annals of Physics*, 1908.

Mitra, A., P.K. Karmakar, and A.K. Sen. A fresh consideration for evaluating mean atmospheric temperature. *Indian J. Phys.*, 74b(5), 379–382, 2000.

Norbury, J.R., and W.J. White. A rapid response rain gauge. *J. Phys. E*, 4, 601–602, 1971.

Ojo, A., B. Shah, T. Janowski, and M. Shareef. *A toolkit for strategic IT planning in public organizations*. e-Macao Program, Macao, SAR, 2008.

Pan, Q.W., and J.E. Allnutt. 12 GHz fade duration and intervals in the tropics. *IEEE Trans. Antennas Propagation*, 52(3), 693–701, 2004.

Pan, Q.W., G.H. Bryant, J. McMohan, J.E. Allnutt, and F. Haidara. High elevation angle satellite-to-earth 12 GHz propagation measurements in the tropics. *Int. J. Satellite Commun. Networking*, 19(4), 363–384, 2001.

Ramchandran, V., and V. Kumar. Invariance of accumulation time factor of Ku band signals in the tropics. *J. Electromagnetic Waves Appl.*, 19(11), 1501–1509, 2005.

Ramchandran, V., and V. Kumar. Modified attenuation model for tropical regions for Ku band signals. *Int. J. Satellite Commun. Networking*, 25(1), 53–67, 2006.

Ramchandran, V., and V. Kumar. Modified rain attenuation model for tropical regions for Ku band signals. *Int. J. Satellite Commun.*, 25, 53–67, 2007.

Schumacher, C., and R.A. Houze. Stratoform rain in the tropics as seen by TRMM precipitation radar. *Am. Meteorol. Soc.*, 16, 1739–1756, 2003.

Semplek, R.A., and R.H. Turrin. Some measurements of attenuation by rainfall at 18.5 GHz. *Bell Syst. Tech. J.*, 48, 1767, 1969.

Sen, A.K., P.K. Karmakar, A. Mitra, A.K. Devgupta, M.K. Dasgupta, O.P.N. Calla, and S.S. Rana. Radiometric studies of clear air attenuation and atmospheric water vapour at 22.235 GHz. *Atmos. Environ.*, 24A(7), 1909–1913, 1990.

Timothi, K.I., S. Sharma, M. Devi, and A.K. Barbara. Model for estimating rain attenuation at frequencies in range 6–30 GHz. *Electron. Lett.*, 31, 1505–1506, 1995.

Ulaby, F.T., R.K. Moore, and A.K. Fung. *Microwave remote sensing: Active and passive: From theory to application.* Vol. 3. Artech House, Norwood, MA, 1986.

Verma, A.K., and K.K. Jha. Rain drop size distribution model for Indian climate. *Indian J. Radio Space Phys.*, 25, 15–21, 1996.

Wu, S.C. Optimum frequencies of a passive radiometer for tropospheric path length correction. *IEEE Trans Antenna Propagation*, AP-27, 233–239, 1979.

Yuter, S.E., and R.A. Houze Jr. Three dimensional kinematic and microphysical evolution of Florida cumulonimbus. Part III. Vertical mass transport, mass divergence and synthesis. *Monthly Weather Rev.*, 123, 1964–1983, 1995.

Zavody, A.M. Effect of scattering by rain on radiometer measurements at millimeter wavelength, PROC. 1EE, 1974. http:// augerlal.lal.in2p3.fr/pmwiki/uploads/Bucholtz.pdf.

Index